Principles of Inorganic Chemistry

Principles of Inorganic Chemistry

Dennis Close

Larsen & Keller
www.larsen-keller.com

Principles of Inorganic Chemistry
Dennis Close
ISBN: 978-1-64172-122-6 (Hardback)

© 2019 Larsen & Keller

 Larsen & Keller

Published by Larsen and Keller Education,
5 Penn Plaza,
19th Floor,
New York, NY 10001, USA

Cataloging-in-Publication Data

Principles of inorganic chemistry / Dennis Close.
    p. cm.
Includes bibliographical references and index.
ISBN 978-1-64172-122-6
1. Chemistry, Inorganic. 2. Chemistry. I. Close, Dennis.
QD151.3 .P75 2019
546--dc23

For more information regarding Larsen and Keller Education and its products, please visit the publisher's website www.larsen-keller.com

Table of Contents

Preface

The synthesis and behavior of organometallic and inorganic compounds are studied in inorganic chemistry. All chemical compounds that do not have a carbon-hydrogen bond are known as inorganic compounds. These are generally classified as coordination compounds, transition metal compounds, cluster compounds, bioinorganic compounds, etc. The concepts of the Bohr model of the atom, ligand field theory, molecular orbital theory, density functional theory, VSEPR theory and the molecular symmetry group theory are integral to the development of this field. Inorganic chemistry has applications in all aspects of the chemical industry, such as in catalysis, coatings, surfactants, pigments, etc. besides the agriculture and medicine industry. This textbook is a valuable compilation of topics, ranging from the basic to the most complex theories and principles in the field of inorganic chemistry. It attempts to understand the multiple branches that fall under this discipline and how such concepts have practical applications. It aims to serve as a resource guide for students and experts alike and contribute to the growth of the discipline.

To facilitate a deeper understanding of the contents of this book a short introduction of every chapter is written below:

Chapter 1, Inorganic chemistry is a field of chemistry that is concerned with the behavior and synthesis of organometallic and inorganic compounds. This chapter has been carefully written to provide an introduction to inorganic chemistry and discusses the essentials of inorganic compounds. **Chapter 2**, An inorganic compound is a chemical compound that does not have C-H bonds. They comprise most of the Earth's crust. The topics elucidated in this chapter cover some of the important inorganic compounds, such as aluminium chloride, ammonia, aluminium hydroxide, barium chlorate, zirconium tungstate, etc. **Chapter 3**, Inorganic chemical reactions generally fall into the broad categories of combination, decomposition, single displacement and double displacement reactions. This chapter closely examines these crucial inorganic reactions. **Chapter 4**, Bioinorganic chemistry delves into the role of metals in the field of biology. It involves the study of natural phenomena like the behavior of metalloproteins and artificially introduced metals in medicine and toxicology. An elaborate study of the varied principles of bioinorganic chemistry has been provided in this chapter. **Chapter 5**, Some of the significant theories central to the development of inorganic chemistry include the crystal field theory, ligand field theory and the theory of molecular symmetry. This chapter discusses in extensive detail about these theories of inorganic chemistry. **Chapter 6**, The bulk properties of inorganic compounds, their optical and electronic properties are studied using diverse techniques. Some common techniques used are spectroscopy, X-ray crystallography, electrochemistry and dual polarization interferometry, among others. All such important techniques for the characterization of inorganic compounds have been covered in this chapter. **Chapter 7**, A coordination compound consists of a central metallic atom or ion surrounded by an array of ions or bound molecules. The study of coordination compounds is under the science of coordination chemistry. Some of the diverse topics covered in this chapter include coordination compounds, coordination complex, coordination number, trans and cis effect, etc. **Chapter 8**, Organometallic compounds are chemical compounds that contain one or more chemical bonds between a carbon atom and a metal or an organic molecule. The bond may also be between a carbon atom and a transition metal or alkaline or alkaline earth metal. The study of organometallic compounds is studied under organometallic chemistry. This is an important chapter, which will analyze in detail about the principles of organometallic chemistry, organometallic compounds and organometallic reactions.

Finally, I would like to thank the entire team involved in the inception of this book for their valuable time and contribution. This book would not have been possible without their efforts. I would also like to thank my friends and family for their constant support.

Dennis Close

Introduction to Inorganic Chemistry

Inorganic chemistry is a field of chemistry that is concerned with the behavior and synthesis of organometallic and inorganic compounds. This chapter has been carefully written to provide an introduction to inorganic chemistry and discuss the essentials of inorganic compounds.

Inorganic chemistry is the branch of chemistry concerned with investigation of the properties of all elements, and the properties and methods of syntheses of their compounds, except for carbon and most carbon-containing compounds. (The study of some carbon-containing compounds—such as carbon dioxide, carbonates, and cyanides—is considered part of inorganic chemistry.) This field stands in a complementary relation-ship to organic chemistry, which covers the myriad carbon-based compounds. These two disciplines are generally considered separately, but there is much overlap, such as in the sub-discipline of organometallic chemistry.

Important classes of inorganic compounds include the oxides, sulfides, sulfates, carbonates, nitrates, and halides. Many of them are found in inanimate materials, such as minerals. For example, soil may contain iron sulfide as pyrite or calcium sulfate as gypsum. A number of inorganic compounds are found in biological systems, such as in the form of electrolytes (sodium chloride).

The study of inorganic chemistry has led to enormous benefits in practical terms. Traditionally, the scale of a nation's economy could be evaluated by its productivity of sulfuric acid. In 2005, the top 20 inorganic chemicals manufactured in Canada, China, Europe, Japan, and the United States were (in alphabetical order): Aluminum sulfate, ammonia, ammonium nitrate, ammonium sulfate, carbon black, chlorine, hydrochloric acid, hydrogen, hydrogen peroxide, nitric acid, nitrogen, oxygen, phosphoric acid, sodium carbonate, sodium chlorate, sodium hydroxide, sodium silicate, sodium sulfate, sulfuric acid, and titanium dioxide.

Most inorganic compounds occur as salts, in which cations and anions are held together by ionic bonds. Examples of cations are sodium (Na^+) and magnesium (Mg^{2+}) examples of anions are oxide (O^{2-}) and chloride (Cl^-). These ions form compounds such as sodium oxide (Na_2O) or magnesium chloride ($MgCl_2$), which are neutral in their overall charge. The ions are described by their oxidation state and their ease of formation can be inferred from the ionization potential (for cations) or from the electron affinity (for anions) of the parent elements.

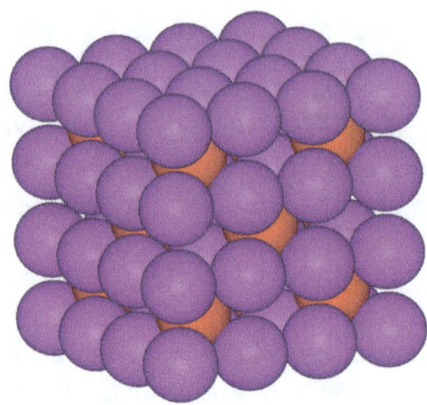

Many inorganic compounds are characterized by high melting points. Inorganic salts typically are poor conductors in the solid state. Other characteristic properties of inorganic compounds are their solubility in water (and other solvents) and ease of crystallization. Some compounds (such as sodium chloride, $NaCl$) are very soluble in water, others (such as silicon dioxide, SiO_2) are not.

A simple inorganic reaction is double displacement, in which the ions of two salts are swapped without a change in oxidation state. In redox reactions, the oxidation state of one reactant, the oxidant, decreases, and that of the other reactant, the reductant, increases. The net result is an exchange of electrons. Electron exchange can occur indirectly as well, such as in electrical batteries—a key feature in electrochemistry.

Some inorganic compounds are acids or bases, and they undergo acid-base reactions. By the Brønsted-Lowry definition, an acid is a proton (hydrogen ion) donor; a base is a proton acceptor. By the Lewis definition, which is more general, any chemical species capable of binding to an electron pair is called a Lewis acid; conversely, any molecule that tends to donate an electron pair (to form a bond) is called a Lewis base.

The first important man-made inorganic compound was ammonium nitrite for soil fertilization through the Haber process. Inorganic compounds are synthesized for use as catalysts such as vanadium(V) oxide and titanium(III) chloride, or as reagents in organic chemistry such as lithium aluminum hydride.

Subdivisions of inorganic chemistry are organometallic chemistry, cluster chemistry, and bioinorganic chemistry. These fields are active areas research in inorganic chemistry, aimed toward new catalysts, superconductors, and therapies.

Descriptive Inorganic Chemistry

Descriptive inorganic chemistry focuses on the classification of compounds based on their properties. Partly the classification focuses on the position in the periodic table of the heaviest element (the element with the highest atomic weight) in the compound, partly by grouping compounds by their structural similarities. When studying inorganic compounds, one often encounters parts of the different classes of inorganic chemistry (an organometallic compound is characterized by its coordination chemistry, and may show interesting solid state properties).

Different classifications are:

Coordination Compounds

Classical coordination compounds feature metals bound to "lone pairs" of electrons residing on the main group atoms of ligands such as H_2O, NH_3, Cl^-, and CN^-. In modern coordination, compounds almost all organic and inorganic compounds can be used as ligands. The "metal" usually is a metal from the groups 3-13, as well as the trans-lanthanides and trans-actinides, but from a certain perspective, all chemical compounds can be described as coordination complexes.

The stereochemistry of coordination complexes can be quite rich, as hinted at by Werner's separation of two enantiomers of $\left[Co\left((OH)_2 Co(NH_3)_4 \right)_3 \right]^{6+}$, an early demonstration that chirality is not inherent to organic compounds. A topical theme within this specialization is supramolecular coordination chemistry.

Examples: $\left[Co(EDTA) \right]^-$, $\left[Co(NH_3)_6 \right]^{3+}$, $TiCl_4(THF)_2$.

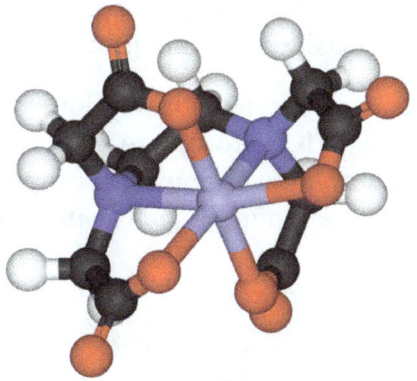

Main Group Compounds

These species feature elements from groups 1, 2 and 13-18 (excluding hydrogen) of the periodic table. Due to their often similar reactivity, the elements in group 3 (Sc, Y, and La) and group 12 (Zn, Cd, and Hg) are also generally included.

Main group compounds have been known since the beginnings of chemistry, for example, elemental sulfur and the distillable white phosphorus. Experiments on oxygen, O_2, by Lavoisier and Priestley not only identified an important diatomic gas, but opened the way for describing compounds and reactions according to stoichiometric ratios. The discovery of a practical synthesis of ammonia using iron catalysts by Carl Bosch and Fritz Haber in the early 1900s deeply impacted mankind, demonstrating the significance of inorganic chemical synthesis. Typical main group compounds are SiO_2, $SnCl_4$, and N_2O.. Many main group compounds can also be classed as "organometallic," as they contain organic groups, for example, $(CH_3)_3)$. Main group compounds also occur in nature, for example, phosphate in DNA, and therefore may be classed as bioinorganic. Conversely, organic compounds lacking (many) hydrogen ligands can be classed as "inorganic," such as the fullerenes, buckytubes and binary carbon oxides.

Examples: tetrasulfur tetranitride S_4N_4,, diborane B_2H_6, silicones, buckminsterfullerene C_{60}.

Transition Metal Compounds

Compounds containing metals from group 4 to 11 are considered transition metal compounds. Compounds with a metal from group 3 or 12 are sometimes also incorporated into this group, but also often classified as main group compounds.

Transition metal compounds show a rich coordination chemistry, varying from tetrahedral for titanium (for example, TiCl4) to square planar for some nickel complexes to octahedral for coordination complexes of cobalt. A range of transition metals can be found in biologically important compounds, such as iron in hemoglobin.

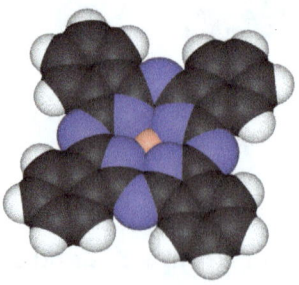

Examples: iron pentacarbonyl, titanium tetrachloride, cisplatin

Organometallic Compounds

Usually, organometallic compounds are considered to contain the M-C-H group. The metal (M) in these species can either be a main group element or a transition metal. Operationally, the definition of an organometallic compound is more relaxed to include also highly lipophilic complexes such as metal carbonyls and even metal alkoxides.

Organometallic compounds are mainly considered a special category because organic ligands are often sensitive to hydrolysis or oxidation, necessitating that organometallic chemistry employs more specialized preparative methods than was traditional in Werner-type complexes. Synthetic methodology, especially the ability to manipulate complexes in solvents of low coordinating power, enabled the exploration of very weakly coordinating ligands such as hydrocarbons, H_2, and N_2. Because the ligands are petrochemicals in some sense, the area of organometallic chemistry has greatly benefited from its relevance to industry.

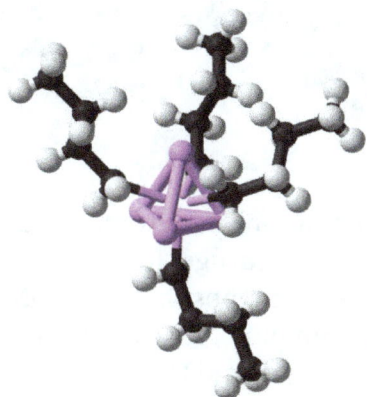

Examples: Cyclopentadienyliron dicarbonyl dimer,
$(C_5H_5)Fe(CO)_2 CH_3$, Ferrocene $Fe(C_5H_5)_2$ Molybdenum hexacarbonyl $Mo(CO)_6$, Diborane B_2H_6, Tetrakis(triphenylphosphine)palladium(0) $Pd[P(C_6H_5)_3]_4$

Cluster Compounds

Clusters can be found in all classes of chemical compounds. According to the commonly accepted definition, a cluster consists minimally of a triangular set of atoms that are directly bonded to each other. But metal-metal bonded dimetallic complexes are highly relevant to the area. Clusters occur in "pure" inorganic systems, organometallic chemistry, main group chemistry, and bioinorganic chemistry. The distinction between very large clusters and bulk solids is increasingly blurred. This interface is the chemical basis of nanoscience or nanotechnology and specifically arise from the study of quantum size effects in cadmium selenide clusters. Thus, large clusters can be described as an array of bound atoms intermediate in character between a molecule and a solid.

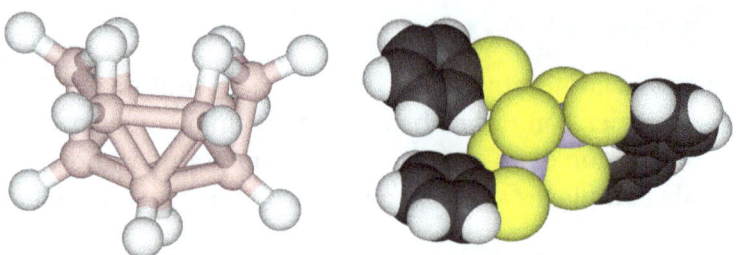

Examples : $Fe_3(CO)_{12}$, $B_{10}H_{14}$,$[Mo_6Cl_{14}]^{2-}$, $4Fe$-$4S$

Bioinorganic Compounds

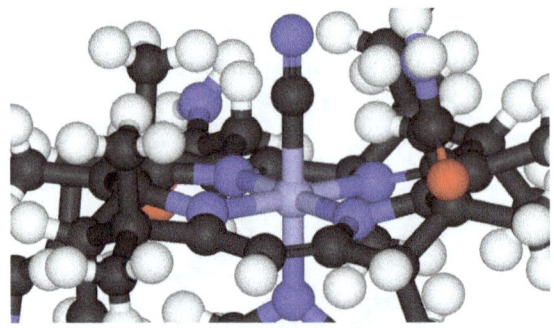

These compounds occur (by definition) in nature, but the subfield includes anthropogenic species, such as pollutants and drugs, for example, Cisplatin. The field includes many compounds, for example, the phosphates in DNA, but also metal complexes containing ligands that range from biological macromolecules, commonly peptides, to ill-defined species such as humic acid, and to water (for example, coordinated to gadolinium complexes employed for MRI).

Solid State Compounds

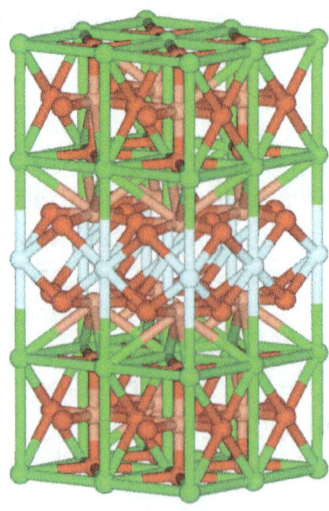

This important area focuses on structure, bonding, and the physical properties of materials. In practice, solid state inorganic chemistry uses techniques such as crystallography to gain an understanding of the properties that result from collective interactions between the subunits of the solid. Included in solid state chemistry are metals and their alloys or intermetallic derivatives. Related fields are condensed matter physics, mineralogy, and materials science.

Examples : silicon chips, zeolites, $YBa_2Cu_3O_7$

Theoretical Inorganic Chemistry

An alternative perspective on the area of inorganic chemistry begins with the Bohr model of the atom and, using the tools and models of theoretical chemistry and computational chemistry, expands into bonding in simple and then more complex molecules. Precise quantum mechanical descriptions for multielectron species, the province of inorganic chemistry, is difficult. This challenge has spawned many semi-quantitative or semi-empirical approaches including molecular orbital theory and ligand field theory, In parallel with these theoretical descriptions, approximate methodologies are employed, including density functional theory.

Exceptions to theories, qualitative and quantitative, are extremely important in the development of the field. For example, $Cu^{II}_2(OAc)_4(H_2O)_2$ is almost diamagnetic below room temperature whereas Crystal Field Theory predicts that the molecule would have two unpaired electrons. The disagreement between qualitative theory (paramagnetic) and observation (diamagnetic) led to the development of models for "magnetic coupling." These improved models led to the development of new magnetic materials and new technologies.

Qualitative Theories

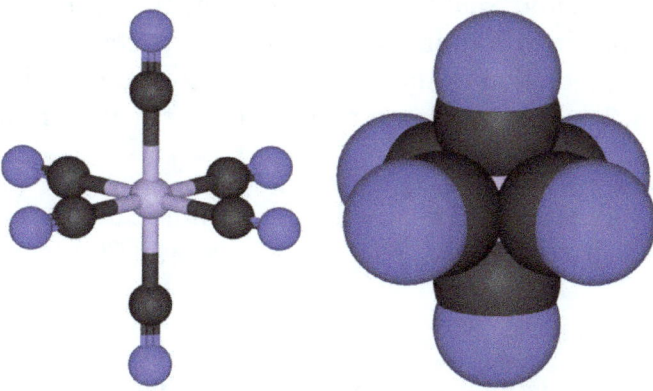

Inorganic chemistry has greatly benefited from qualitative theories. Such theories are easier to learn as they require little background in quantum theory. Within main group compounds, VSEPR theory powerfully predicts, or at least rationalizes, the structures of main group compounds, such as an explanation for why NH3 is pyramidal whereas ClF3 is T-shaped. For the transition metals, crystal field theory allows one to understand the magnetism of many simple complexes, such as why $[Fe^{III}(CN)_6]^{3-}$ has only one unpaired electron, whereas $[Fe^{III}(H_2O)_6]^{3+}$ has five. A particularly powerful qualitative approach to assessing the structure and reactivity begins with classifying molecules according to electron counting, focusing on the numbers of valence electrons, usually at the central atom in a molecule.

Group Theory

A central construct in inorganic chemistry is Group Theory. Group Theory provides the language to describe the shapes of molecules according to their "point group symmetry." Group Theory also enables factoring and simplification of theoretical calculations.

Spectroscopic features are analyzed and described with respect to the symmetry properties of the, inter alia, vibrational or electronic states. Knowledge of the symmetry properties of the ground and excited states allows one to predict the numbers and intensities of absorptions in vibrational and electronic spectra. A classic application of Group Theory is the prediction of the number of C-O vibrations in substituted metal carbonyl complexes. The most common applications of symmetry to spectroscopy involve vibrational and electronic spectra.

As an instructional tool, Group Theory highlights commonalities and differences in the bonding otherwise disparate species, such as WF_6 and $Mo(CO)_6$ or CO_2 and NO_2.

Reaction Pathways

The theory of chemical reactions is more challenging than the theory for a static molecule. Marcus theory provides a powerful linkage between bonding, mechanism, and reactivity. The relative strengths of metal-ligand bonds, which can be calculated theoretically, anticipate the kinetically accessible pathways.

Thermodynamics and Inorganic Chemistry

An alternative quantitative approach to inorganic chemistry focuses on energies of reactions. This approach is highly traditional and empirical, but it is also useful. Broad concepts that are couched in thermodynamic terms include redox potential, acidity, phase changes. A classic concept in inorganic thermodynamics is the Born-Haber cycle, which is used for assessing the energies of elementary processes such as electron affinity, some of which cannot be observed directly.

Mechanistic Inorganic Chemistry

An important and increasingly popular aspect of inorganic chemistry focuses on reaction pathways. The mechanisms of reactions are discussed differently for different classes of compounds.

Main Group Elements and Lanthanides

The mechanisms of main group compounds of groups 13-18 are usually discussed in the context of organic chemistry (organic compounds are main group compounds, after all). Elements heavier than C, N, O, and F often form compounds with more electrons than predicted by the octet rule. The mechanisms of their reactions differ from organic compounds for this reason. Elements lighter than carbon (B, Be, Li) as well as Al and Mg often form electron-deficient structures that are electronically akin to carbocations. Such electron-deficient species tend to react via associative pathways. The chemistry of the lanthanides mirrors many aspects of chemistry seen for aluminium.

Transition Metal Complexes

Mechanisms for the reactions of transition metals are discussed differently from main group compounds. The important role of d-orbitals in bonding strongly influences the pathways and rates of ligand substitution and dissociation.

An overarching aspect of mechanistic transition metal chemistry is the kinetic lability of the complex illustrated by the exchange of free and bound water in the prototypical complexes $[M(H_2O)_6]^{n+}$:

$$[M(H_2O)_6]^{n+} + 6\,H_2O^* \longrightarrow [M(H_2O^*)_6]^{n+} + 6\,H_2O$$

where H_2O^* denotes isotopically enriched water, e.g. $H_2^{17}O$

The rates of water exchange varies by 20 orders of magnitude across the periodic table, with lanthanide complexes at one extreme and Ir(III) species being the slowest.

Redox Reactions

Redox reactions are prevalent for the transition elements. Two classes of redox reaction are considered: Atom-transfer reactions, such as oxidative addition/reductive elimination, and electron-transfer. A fundamental redox reaction is "self-exchange," which involves the degenerate reaction between an oxidant and a reductant. For example, permanganate and its one-electron reduced relative manganate exchange one electron:

$$[MnO_4]^- + [Mn^*O_4]^{2-} \longrightarrow [MnO_4]^{2-} + [Mn^*O_4]^-$$

Reactions at Ligands

Coordinated ligands display reactivity distinct from the free ligands. For example, the acidity of the ammonia ligands in $\left[Co(NH_3)_6\right]^{3+}$ is elevated relative to NH3 itself. Alkenes bound to metal cations are reactive toward nucleophiles whereas alkenes normally are not. The large and industrially important area of catalysis hinges on the ability of metals to modify the reactivity of organic ligands. Homogeneous catalysis occurs in solution and heterogeneous catalysis occurs when gaseous or dissolved substrates interact with surfaces of solids. Traditionally homogeneous catalysis is considered part of organometallic chemistry and heterogeneous catalysis is discussed in the context of surface science, a subfield of solid state chemistry. But the basic inorganic chemical principles are the same. Transition metals, almost uniquely, react with small molecules such as CO, H_2, O_2, and C_2H_4.. The industrial significance of these feedstocks drives the active area of catalysis.

Characterization of Inorganic Compounds

Because of the diverse range of elements and the correspondingly diverse properties of the resulting derivatives, inorganic chemistry is closely associated with many methods of analysis. Older methods tended to examine bulk properties such as the electrical conductivity of solutions, melting points, solubility, and acidity. With the advent of quantum theory and the corresponding expansion of electronic apparatus, new tools have been introduced to probe the electronic properties of inorganic molecules and solids. Often these measurements provide insights relevant to theoretical models. For example, measurements on the photoelectron spectrum of methane demonstrated that describing the bonding by the two-center, two-electron bonds predicted between the carbon and hydrogen using Valence Bond Theory is not appropriate for describing ionization processes in a simple way. Such insights led to the popularization of molecular orbital theory as fully delocalized orbitals are a more appropriate simple description of electron removal and electron excitation.

Commonly encountered techniques are:

- X-ray crystallography: This technique allows for the 3D determination of molecular structures.

- Spectroscopy: Historically, UV-vis spectroscopy has been an important tool, since many inorganic compounds are strongly colored.

- Electron-spin resonance: ESR (or EPR) allows for the measurement of the environment of paramagnetic metal centra.

- Electrochemistry: Cyclic voltammetry and related techniques probe the redox characteristics of compounds.

- NMR spectroscopy: Besides 1H and 13C many other "good" NMR nuclei (for example, ^{11}B, ^{19}F, ^{31}P, and ^{195}Pt) give important information on compound properties and structure. Also the NMR of paramagnetic species can result in important structural information.

- Electron-nuclear double resonance (ENDOR) spectroscopy.

- Mössbauer spectroscopy.

Synthetic Inorganic Chemistry

Although some inorganic species can be obtained in pure form from nature, most are synthesized in chemical plants and in the laboratory.

Inorganic synthetic methods can be classified roughly according the volatility or solubility of the component reactants. Soluble inorganic compounds are prepared using methods of organic synthesis. For metal-containing compounds that are reactive toward air, Schlenk line and glove box techniques are followed. Volatile compounds and gases are manipulated in "vacuum manifolds" consisting of glass piping interconnected through valves, the entirety of which can be evacuated to 0.001 mm Hg or less. Compounds are condensed using liquid nitrogen (b.p. 78K) or other cryogens. Solids are typically prepared using tube furnaces, the reactants and products being sealed in containers, often made of fused silica (amorphous $SiO2$) but sometimes more specialized materials such as welded Ta tubes or Pt "boats." Products and reactants are transported between temperature zones to drive reactions.

Types of Reactions and Examples

There are about four types chemical reactions of Inorganic chemistry namely combination, decomposition, single displacement and double displacement reactions.

Combination Reactions

As it is in the name 'Combination', here two or more substances combine to form a product which is called as Combination reaction. For example:

$$Barium + F_2 \longrightarrow BaF_2$$

Decomposition Reaction

It is a type of reaction where a single element splits up or say decomposes into two products. For example:

$$FeS \longrightarrow Fe + S$$

Single Displacement Reactions

A reaction where a single atom of one element replaces another atom of one more element. For example:

$$Zn(s) + CuSO_4(aq) \longrightarrow Cu(s) + ZnSO_4(aq)$$

Double Displacement Reactions

This type of reaction is also called as 'metathesis reactions'. Here two elements of two different compounds displace each other to form two new compounds. For example:

$$CaCl_2(aq) + 2AgNO_3(aq) \longrightarrow Ca(NO_3)_2(aq) + 2AgCl(s)$$

Applications of Inorganic Chemistry

Inorganic chemistry finds its high number of applications in various fields such as Biology, chemical, engineering, etc.

- It is applied in the field of medicine and also in healthcare facilities.
- The most common application is the use of common salt or the compound Sodium hydroxide in our daily lives.
- Baking soda is used in the preparation of cakes and other foodstuffs.
- Many inorganic compounds are utilized in ceramic industries.
- In the electrical field, it is applied to the electric circuits as silicon in the computers, etc.

Inorganic Compounds

An inorganic compound is any compound that lacks a carbon atom, for lack of a more in-depth definition. Those compounds with a carbon atom are called organic compounds, due to their root base in an atom that is vital for life. There are a small number of inorganic compounds that actually do contain carbon, given its propensity for forming molecular bonds; these include carbon monoxide and carbon dioxide, to name a few.

Inorganic compounds are often quite simple, as they do not form the complex molecular bonds that carbon makes possible. A common example of a simple inorganic compound would be sodium chloride, known more commonly as household salt. This compound contains only two atoms, sodium (Na) and chlorine (Cl).

Examples of Inorganic Compounds:

1. H_2O - Water is a simple inorganic compound, even though it contains hydrogen, a key atom (along with carbon) in many organic compounds. The atoms in a molecule of water have formed very simple bonds due to this lack of carbon.

2. HCl - Hydrochloride, also known as hydrochloric acid when it is dissolved in water, is a colorless, corrosive acid with a fairly strong pH. It is found in the gastric juices of many animals, helping in digestion by breaking down food.

3. CO_2 - Carbon dioxide, despite the presence of a carbon atom in the formula, is classified as an inorganic compound. This has caused a dispute within the scientific community, with questions being raised as to the validity of our current methods of classifying compounds. Currently, organic compounds contain a carbon or a hydrocarbon, which forms a stronger bond. The bond formed by carbon in CO_2 is not a strong bond.

4. NO_2 - Nitrogen dioxide gas presents a variety of colors at different temperatures. It is often produced in atmospheric nuclear tests, and is responsible for the tell-tale reddish color displayed in mushroom clouds. It is highly toxic, and forms fairly weak bonds between the nitrogen and oxygen atoms.

5. Fe_2O_3 - Iron (III) oxide is one of the three main oxides of iron, and is an inorganic compound due to the lack of a carbon atom or a hydrocarbon. Iron (III) oxide occurs naturally as hematite, and is the source of most iron for the steel production industry. It is commonly known as rust, and shares a number of characteristics with its naturally occurring counterpart.

2

Inorganic Compounds

An inorganic compound is a chemical compound that does not have C-H bonds. They comprise most of the Earth's crust. The topics elucidated in this chapter cover some of the important inorganic compounds, such as aluminium chloride, ammonia, aluminium hydroxide, barium chlorate, zirconium tungstate, etc.

Inorganic compound can be considered as a compound that does not contain a carbon-to-hydrogen bond, also called a C-H bond. Moreover, inorganic compounds tend to be minerals or geologically-based compounds that do not contain carbon-to-hydrogen bonds. Not all, but most inorganic compounds contain a metal. That said, there are countless compounds that fall under the realm of inorganic. In fact, the majority of all compounds in this universe are inorganic in nature. For this reason, inorganic compounds have an overwhelming amount of applications and practical uses in the real world. Since most of the compounds in this world are inorganic, these compounds can take on a host of forms and possess many different characteristics.

Names and Formulas of Inorganic Compounds

Many compounds, particularly those that have been known for a relatively long time, have more than one name: a common name (sometimes several), and a systematic name, which is the name assigned by adhering to specific rules. Like the names of most elements, the common names of chemical compounds generally have historical origins, although they often appear to be unrelated to the compounds of interest. For example, the systematic name for KNO_3 is potassium nitrate, but its common name is saltpeter.

a systematic nomenclature is used to assign meaningful names to the millions of known substances. Unfortunately, some chemicals that are widely used in commerce and industry are still known almost exclusively by their common names; in such cases, familiarity with the common name as well as the systematic one is required.

Binary ionic compounds contain only two elements. The procedure for naming such compounds is outlined in Figure below and uses the following steps:

1. Place the ions in their proper order: cation and then anion.

2. Name the cation.

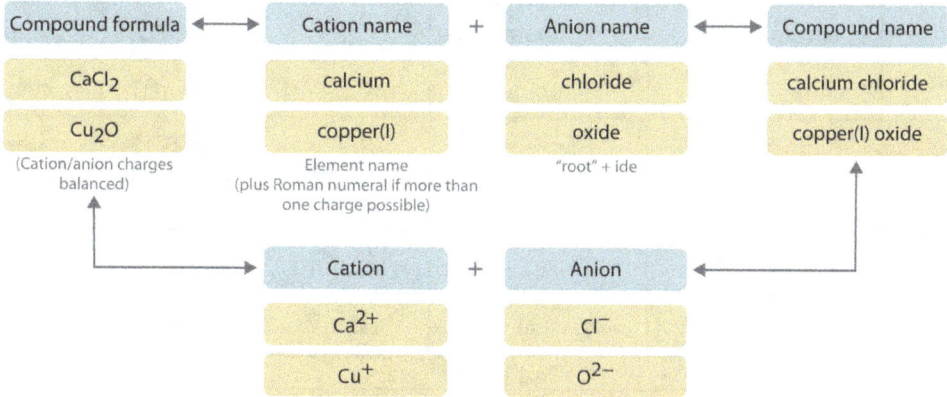

Naming an Ionic Compound

a. Metals that form only one cation, these metals are usually in Groups 1–3, 12, and 13. The name of the cation of a metal that forms only one cation is the same as the name of the metal (with the word ion added if the cation is by itself). For example, Na^+ is the sodium ion, Ca^{2+} is the calcium ion, and Al^{3+} is the aluminum ion.

b. Metals that form more than one cation. Many metals can form more than one cation. This behavior is observed for most transition metals, many actinides, and the heaviest elements of Groups 13–15. In such cases, the positive charge on the metal is indicated by a roman numeral in parentheses immediately following the name of the metal. Thus Cu^+ is copper(I) (read as "copper one"), Fe^{2+} is iron(II), Fe^{3+} is iron(III), Sn^{2+} is tin(II), and Sn^{4+} is tin(IV).

An older system of nomenclature for such cations is still widely used, however-er. The name of the cation with the higher charge is formed from the root of the element's Latin name with the suffix -ic attached, and the name of the cat-ion with the lower charge has the same root with the suffix -ous. The names of Fe^{3+}, Fe^{2+}, Sn^{4+}, and Sn^{2+} are therefore ferric, ferrous, stannic, and stannous, respectively. Even though this text uses the systematic names with roman nu-merals, it is important to recognize these common names because they are still often used. For example, on the label of dental fluoride rinse, the compound chemists call tin(II) fluoride is usually listed as stannous fluoride.

Some examples of metals that form more than one cation are listed in table, along with the names of the ions. Note that the simple Hg+ cation does not occur in chemical compounds. Instead, all compounds of mercury(I) contain a dimeric cation, Hg_2^{2+}, in which the two Hg atoms are bonded together.

Cation	Systematic Name	Common Name	Cation	Systematic Name	Common Name
Co^{2+}	cobalt(II)	cobaltous*	Pb^{4+}	lead(IV)	plumbic*
Co^{3+}	cobalt(III)	cobaltic*	Pb^{2+}	lead(II)	plumbous*

Cr^{2+}	chromium(II)	chromous	Cu^{2+}	copper(II)	cupric
Cr^{3+}	chromium(III)	chromic	Cu^{+}	copper(I)	cuprous
Fe^{2+}	iron(II)	ferrous	Sn^{4+}	tin(IV)	stannic
Fe^{3+}	iron(III)	ferric	Sn^{2+}	tin(II)	stannous
Mn^{2+}	manganese(II)	manganous*	Hg^{2+}	mercury(II)	mercuric
Mn^{3+}	manganese(III)	manganic*	Hg_2^{2+}	mercury(I)	mercurous†
* Not widely used.					
†The isolated mercury(I) ion exists only as the gaseous ion.					

    c.   Polyatomic cations. The names of the common polyatomic cations that are relatively important in ionic compounds (such as, the ammonium ion) are in Table above "Common Polyatomic Ions and Their Names."

3. Name the anion.

    a.   Monatomic anions. Monatomic anions are named by adding the suffix -ide to the root of the name of the parent element; thus, Cl⁻ is chloride, O^{2-} is oxide, P^{3-} is phosphide, N^{3-} is nitride (also called azide), and C^{4-} is carbide. Because the charges on these ions can be predicted from their position in the periodic table, it is not necessary to specify the charge in the name. Examples of monatomic anions are in table "Some Common Monatomic Ions and Their Names."

    b.   Polyatomic anions. Polyatomic anions typically have common names that must be memorized; some examples are in table. Polyatomic anions that contain a single metal or nonmetal atom plus one or more oxygen atoms are called oxoanions (or oxyanions). In cases where only two oxoanions are known for an element, the name of the oxoanion with more oxygen atoms ends in -ate, and the name of the oxoanion with fewer oxygen atoms ends in -ite. For example, NO_3^- is nitrate and NO_2^- is nitrite.

The halogens and some of the transition metals form more extensive series of oxoanions with as many as four members. In the names of these oxoanions, the prefix per- is used to identify the oxoanion with the most oxygen (so that ClO_4^- is perchlorate and ClO_3^- is chlorate), and the prefix hypo- is used to identify the anion with the fewest oxygen (ClO_2^- is chlorite and ClO^- is hypochlorite). The relationship between the names of oxoanions and the number of oxygen atoms present is diagrammed in figure. "The Relationship between the Names of Oxoanions and the Number of Oxygen Atoms Present." Differentiating the oxoanions in such a series is no trivial matter; for example, the hypochlorite ion is the active ingredient in laundry bleach and swimming pool disinfectant, but compounds that contain the perchlorate ion can explode if they come into contact with organic substances.

4. Write the name of the compound as the name of the cation followed by the name of the anion.

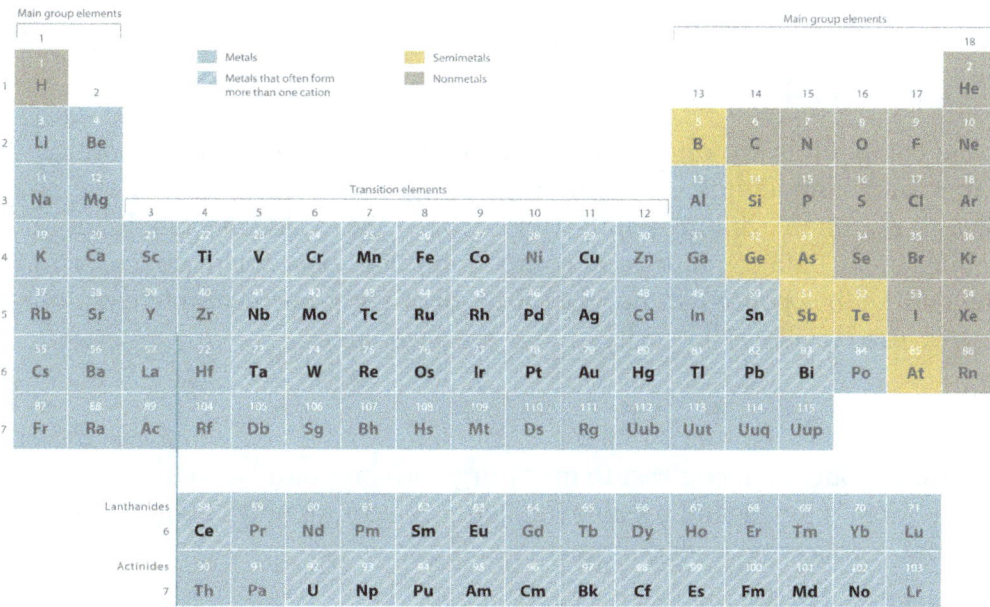

It is not necessary to indicate the number of cations or anions present per formula unit in the name of an ionic compound because this information is implied by the charges on the ions. The charge of the ions must be considered when writing the formula for an ionic compound from its name, however. Because the charge on the chloride ion is −1 and the charge on the calcium ion is +2, for example, consistent with their positions in the periodic table, arithmetic indicates that calcium chloride must contain twice as many chloride ions as calcium ions to maintain electrical neutrality. Thus, the formula is $CaCl_2$. Similarly, calcium phosphate must be $Ca_3(PO_4)_2$ because the cation and the anion have charges of +2 and −3, respectively. The best way to learn how to name ionic compounds is to work through a few examples, referring to figure and tables as needed.

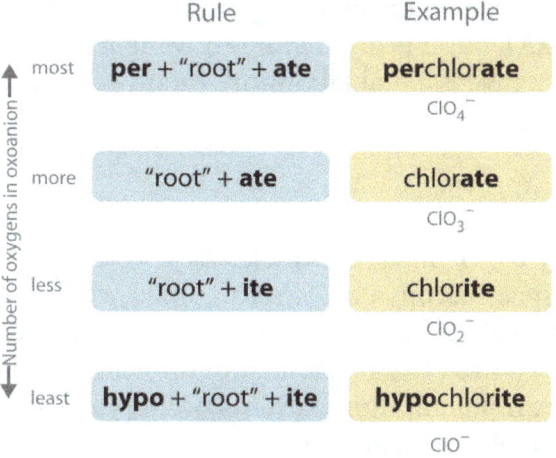

The Relationship between the Names of Oxoanions and the Number of Oxygen Atoms Present.

Physical Characteristics

Melting/Boiling Points

Inorganic compounds are often ionic, and so have very high melting points. While some inorganic compounds are solids with accessible melting points, and some are liquids with reasonable boiling points, there are not the exhaustive tabulations of melting/boiling point data for inorganic compounds that exist for organics. In general, melting and boiling points are not useful in identifying an inorganic compound, but they can be used to assess its purity, if they are accessible.

Color

Inorganic compounds, in contrast to many organic compounds, are very colorful. Unfortunately, color alone is not reliable indicator of a compound's identity, but it is useful when one is following a separation on a column.

Crystal Shapes

You can get information about the arrangement of the particles in the solid from the shape of a well-formed crystal, or by observing the visual changes in the crystal when it is rotated under a polarizing microscope. Crystal shape is used as a means of mineral identification, but mineralogists have the advantage over chemists in that their crystals are the result of very long, slow crystallization processes that often result in large, well-formed crystals.

Elemental Analysis

Elemental analysis is one of the most useful methods available to characterize a compound. You can do some elemental analyses yourself using standard procedures (gravimetric, colorimetric, AA), but it is often more convenient to pay someone else to do it. For a C, H, N analysis a professional laboratory requires ~10mg of sample and ~ $ 50, but it is definitely worth it.

Mass Spectroscopy

In inorganic chemistry this is most often used to determine the molar mass of compounds. When mass spectroscopy data are combined with an elemental analysis, the chemical formula of the substance can be determined. Analysis of a compound's fragmentation pattern can be used to gain structural information. This is not usually done because of the complex fragmentation patterns of inorganic compounds, and because other methods are available for structure determination.

Chromatography

Most often used to separate a product from a complex reaction mixture. If a known sample of the compound is available, it can be identified in a reaction mixture by spiking the analyte with the known. Most chromatography in inorganic chemistry is on solutions (e. g., column chromatography and HPLC, both normal and reverse-phase), because the high boiling point of many inorganic compounds precludes GC analysis.

Other Physical Properties

In the old days taste and smell were used to characterize compounds. However, these are no longer recommended because of the potential toxicity of new compounds.

A useful physical property that can be used to distinguish between ionic and non-ionic compounds is "crunchiness." When crushed with a spatula most ionic compounds will feel crunchy, while non-ionic compounds will not. This should not be tried with azide, perchlorate or fulminate salts because these tend to be explosive.

Spectroscopic/Structural Methods

UV-Vis Absorption Spectroscopy

The number, energies and intensities of a transition metal compound's absorption bands in the UV-Vis and near IR can be used to determine the general type of atom bound to a metal and the geometry about the metal.

Circular Dichroism (CD) Spectroscopy

CD spectroscopy measures the degree to which a sample rotates circularly polarized light. CD is limited to enantiomerically pure, optically-active compounds; an equal mixture of enantiomers will give no CD spectrum (enantiomeric pairs having CD spectra that are mirror images of each other and so would cancel out). CD is similar to absorption spectroscopy, but because of different selection rules for the electronic transitions, peak intensities and widths may be different between CD and absorption (the energy of a peak must still, however, be the same). This is a very useful technique for studying metal-containing proteins and enzymes because the chirality of the peptide induces a CD signal from any metal bound to the protein.

IR Absorption Spectroscopy

Can be used just like in organic chemistry to fingerprint a compound. In simple compounds the number, energy and intensity of the IR transitions are directly related to the geometry of compound and to which atoms are bound to which other atoms. Unfortunately, some metal-ligand vibrations, and many of the vibrations for the heavier elements occur outside the frequency window of most commercial instruments. For complex compounds involving large organic moieties, the IR becomes more difficult to interpret. IR is useful to determine the presence of complex counter ions like PF_6^-, ClO_4^-, BF_4^- because they have distinctive absorptions in the IR.

Raman Spectroscopy

Raman spectroscopy is complementary to IR absorption spectroscopy. Both probe vibrations within a compound, but they have different selection rules. By considering what peaks are present, or absent, in the two spectra of a compound, one can determine geometry; at least in simple cases. Raman is also useful because it can, depending on instrument design, scan to very low frequencies (~100 cm-1), and thus observe transitions too low for IR absorption. A variant technique, resonance Raman, can be used to assign vibrations and identify ligands.

Nuclear Magnetic Resonance (NMR)

The workhorse of chemistry. While paramagnetic NMR is not impossible, NMR is usually performed on diamagnetic compounds. The NMR spectra of inorganic compounds are often more complicated than organics because other nuclei also have nuclear magnetic moments. So in addition to the familiar ^1H ($I = 1/2$) and ^{13}C ($I = 1/2$), there are ~90 other elements that have at least one NMR-active nucleus. Although less widespread than the standard solution NMR, solid state NMR and even single-crystal NMR have been used on materials that simply do not dissolve in any solvent.

Magnetism

The number of unpaired electrons in a transition metal compound is a very useful physical property to know. From a knowledge of the magnetism one can determine what metal is present, its oxidation state and even a rough idea of the metal's structure. Magnetic measurements are often made with a Gouy balance, an Evans balance (a modified version of the Gouy balance, which we have in lab), or a SQUID (super-quantum interference device) magnetometer. A compound's magnetic moment can also be determined by NMR using the Evans method. Its one drawback is that it is a bulk technique, and when used on mixtures the measured magnetic moment will be the weighted average of the moments of all of the paramagnetic species present.

Electron Paramagnetic Resonance (EPR or ESR)

While NMR is usually only for diamagnetic compounds, EPR is for paramagnetic compounds with an odd number of unpaired electrons (EPR can be done when there are an even number of unpaired electrons, but it is a much harder experiment). In this experiment a sample is irradiated with microwave radiation and the field is swept until resonance occurs. The field at which resonance occurs depends on the number of un-paired electrons, the geometry about the metal center and the metal's ligands. In many ways EPR and NMR the same, and there is even a technique that combines them (EN-DOR, electron-nuclear double resonance, spectroscopy).

Magnetic Circular Dichroism (MCD)

MCD is a hybrid technique based on the fact that all matter will rotate circularly polar-ized light in the presence of a magnetic field. It is most useful for paramagnetic com-pounds (either with an even or odd number of unpaired electrons). Because it is both a magnetic and a spectroscopic method, MCD can be used to measure a compound's magnetic properties and electronic transitions. MCD is most powerful when used in conjunction with another method. It can also be used on a mixture, if transitions aris-ing from different species can be identified. The main limitations of MCD is that it is usually performed at low temperature (< 77 K) and the samples must be strain-free glasses. The cost of the instrumentation and the cryogens have limited MCD to only a few groups worldwide.

Electrochemistry

For most inorganic chemists, electrochemistry means cyclic voltammetry (CV). CV has been called the "electrochemical equivalent of spectroscopy" because it can be used to determine 1) the oxidation state, 2) E0 of each redox process that the compound can undergo and 3) the kinetics of the redox process. CV is a standard characterization method in most inorganic laboratories.

X-Ray Methods

Single-crystal X-ray diffraction is the most powerful X-ray technique for inorganic chemists. From precise measurement of the intensity and angles at which an X-ray beam diffracts off a crystal, the arrangement of the atoms can be reconstructed. Ob-viously, as a direct probe of structure crystallography is an invaluable characteriza-tion method for all types of compounds. Some inorganic compounds (e. g., rocks, and minerals) can't be obtained as single crystals. In these cases X-ray powder diffraction can be used to obtain the dimensions of the unit cell for use in identification (there is a large, indexed catalog of lattice constants for many minerals).

Etc.

There are a large number of techniques for characterizing inorganic compounds. Some are limited to one or two elements (e. g., Mssbauer spectroscopy), while others require large, expensive instruments (e. g., EXAFS). These are not that important in every day work, but you should be aware of them and what their strengths and weak-nesses are.

Inorganic Compounds Essential to Human Functioning

The following four groups of inorganic compounds are essential to life:

As much as 70 percent of an adult's body weight is water. This water is contained both within the cells and between the cells that make up tissues and organs. Its several roles make water indispensable to human functioning.

Water as a Lubricant and Cushion

Water is a major component of many of the body's lubricating fluids. Just as oil lubricates the hinge on a door, water in synovial fluid lubricates the actions of body joints, and water in pleural fluid helps the lungs expand and recoil with breathing. Watery fluids help keep food flowing through the digestive tract, and ensure that the movement of adjacent abdominal organs is friction free.

Water also protects cells and organs from physical trauma, cushioning the brain within the skull, for example, and protecting the delicate nerve tissue of the eyes. Water cushions a developing fetus in the mother's womb as well.

Water as a Heat Sink

A heat sink is a substance or object that absorbs and dissipates heat but does not experience a corresponding increase in temperature. In the body, water absorbs the heat generated by chemical reactions without greatly increasing in temperature. Moreover, when the environmental temperature soars, the water stored in the body helps keep the body cool. This cooling effect happens as warm blood from the body's core flows to the blood vessels just under the skin and is transferred to the environment. At the same time, sweat glands release warm water in sweat. As the water evaporates into the air, it carries away heat, and then the cooler blood from the periphery circulates back to the body core.

Water as a Component of Liquid Mixtures

A mixture is a combination of two or more substances, each of which maintains its own chemical identity. In other words, the constituent substances are not chemically bonded into a new, larger chemical compound. The concept is easy to imagine if you think of powdery substances such as flour and sugar; when you stir them together in a bowl, they obviously do not bond to form a new compound. The room air you breathe is a gaseous mixture, containing three discrete elements—nitrogen, oxygen, and argon—and one compound, carbon dioxide. There are three types of liquid mixtures, all of which contain water as a key component; these are solutions, colloids, and suspensions.

For cells in the body to survive, they must be kept moist in a water-based liquid called a solution. In chemistry, a liquid solution consists of a solvent that dissolves a substance called a solute. An important characteristic of solutions is that they are homogeneous; that is, the solute molecules are distributed evenly throughout the solution. If you were to stir a teaspoon of sugar into a glass of water, the sugar would dissolve into sugar molecules separated by water molecules. The ratio of sugar to water in the left side of the glass would be the same as the ratio of sugar to water in the right side of the glass. If you were to add more sugar, the ratio of sugar to water would change, but the distribution—provided you had stirred well—would still be even.

Water is considered the "universal solvent" and it is believed that life cannot exist without water because of this. Water is certainly the most abundant solvent in the body; essentially all of the body's chemical reactions occur among compounds dissolved in water. Since water molecules are polar, with regions of positive and negative electrical charge, water readily dissolves ionic compounds and polar covalent compounds. Such compounds are referred to as hydrophilic, or "water-loving." As mentioned above, sugar dissolves well in water. This is because sugar molecules contain regions of hydrogen-oxygen polar bonds, making it hydrophilic. Nonpolar molecules, which do not readily dissolve in water, are called hydrophobic, or "water-fearing."

Concentrations of Solutes

Various mixtures of solutes and water are described in chemistry. The concentration of a given solute is the number of particles of that solute in a given space (oxygen makes up about 21 percent of atmospheric air). In the bloodstream of humans, glucose concentration is usually measured in milligram (mg) per deciliter (dL), and in a healthy adult averages about 100 mg/dL. Another method of measuring the concentration of a solute is by its molarilty—which is moles (M) of the molecules per liter (L). The mole of an element is its atomic weight, while a mole of a compound is the sum of the atomic weights of its components, called the molecular weight. An often-used example is calculating a mole of glucose, with the chemical formula $C_6H_{12}O_6$. Using the periodic table, the atomic weight of carbon (C) is 12.011 grams (g), and there are six carbons in glucose, for a total atomic weight of 72.066 g. Doing the same calculations for hydrogen (H) and oxygen (O), the molecular weight equals 180.156g (the "gram molecular weight" of glucose). When water is added to make one liter of solution, you have one mole (1M) of glucose. This is particularly useful in chemistry because of the relationship of moles to "Avogadro's number." A mole of any solution has the same number of particles in it: 6.02×1023. Many substances in the bloodstream and other tissue of the body are measured in thousandths of a mole, or millimoles (mM).

A colloid is a mixture that is somewhat like a heavy solution. The solute particles consist of tiny clumps of molecules large enough to make the liquid mixture opaque (because the particles are large enough to scatter light). Familiar examples of colloids are milk and cream. In the thyroid glands, the thyroid hormone is stored as a thick protein mixture also called a colloid.

A suspension is a liquid mixture in which a heavier substance is suspended temporarily in a liquid, but over time, settles out. This separation of particles from a suspension is called sedimentation. An example of sedimentation occurs in the blood test that establishes sedimentation rate, or sed rate. The test measures how quickly red blood cells in a test tube settle out of the watery portion of blood (known as plasma) over a set period of time. Rapid sedimentation of blood cells does not normally happen in the healthy body, but aspects of certain diseases can cause blood cells to clump together, and these heavy clumps of blood cells settle to the bottom of the test tube more quickly than do normal blood cells.

The Role of Water in Chemical Reactions

Two types of chemical reactions involve the creation or the consumption of water: dehydration synthesis and hydrolysis.

- In dehydration synthesis, one reactant gives up an atom of hydrogen and another reactant gives up a hydroxyl group (OH) in the synthesis of a new product. In the formation of their covalent bond, a molecule of water is released as a by-product ([link]). This is also sometimes referred to as a condensation reaction.

- In hydrolysis, a molecule of water disrupts a compound, breaking its bonds. The water is itself split into H and OH. One portion of the severed compound then bonds with the hydrogen atom, and the other portion bonds with the hydroxyl group.

These reactions are reversible, and play an important role in the chemistry of organic compounds (which will be discussed shortly).

(a) Dehydration synthesis

Monomers are joined by removal of OH from one monomer and removal of H from the other at the site of bond formation.

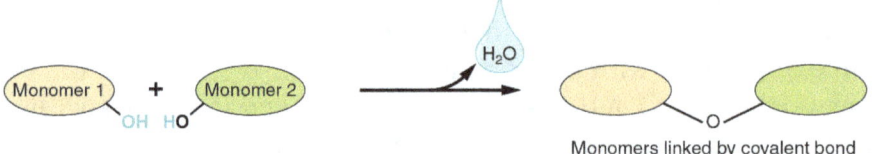

Monomers linked by covalent bond

(b) Hydrolysis

Monomers are released by the addition of a water molecule, adding OH to one monomer and H to the other.

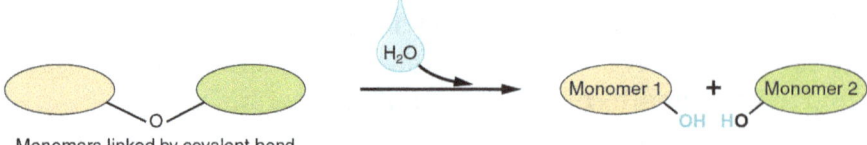

Monomers linked by covalent bond

Monomers, the basic units for building larger molecules, form polymers
(two or more chemically-bonded monomers).

(a) In dehydration synthesis, two monomers are covalently bonded in a reaction in which one gives up a hydroxyl group and the other a hydrogen atom. A molecule of water is released as a byproduct during dehydration reactions. (b) In hydrolysis, the covalent bond between two monomers is split by the addition of a hydrogen atom to one and a hydroxyl group to the other, which requires the contribution of one molecule of water.

Salts

Recall that salts are formed when ions form ionic bonds. In these reactions, one atom gives up one or more electrons, and thus becomes positively charged, whereas the other accepts one or more electrons and becomes negatively charged. You can now define a salt as a substance that, when dissolved in water, dissociates into ions other than H^+ or OH^-. This fact is important in distinguishing salts from acids and bases, discussed next.

A typical salt, NaCl, dissociates completely in water. The positive and negative regions on the water molecule (the hydrogen and oxygen ends respectively) attract the negative chloride and positive sodium ions, pulling them away from each other. Again, whereas nonpolar and polar covalently bonded compounds break apart into molecules in solution, salts dissociate into ions. These ions are electrolytes; they are capable of conducting an electrical current in solution. This property is critical to the function of ions in transmitting nerve impulses and prompting muscle contraction.

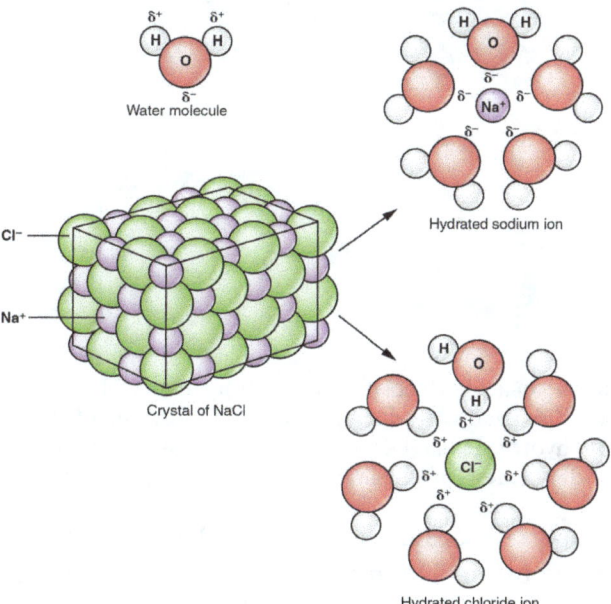

Notice that the crystals of sodium chloride dissociate not into molecules of NaCl, but into Na^+ cations and Cl^- anions, each completely surrounded by water molecules.

Many other salts are important in the body. For example, bile salts produced by the liver help break apart dietary fats, and calcium phosphate salts form the mineral portion of teeth and bones.

Acids and Bases

Acids and bases, like salts, dissociate in water into electrolytes. Acids and bases can very much change the properties of the solutions in which they are dissolved.

Acids

An acid is a substance that releases hydrogen ions (H$^+$) in solution. Because an atom of hydrogen has just one proton and one electron, a positively charged hydrogen ion is simply a proton. This solitary proton is highly likely to participate in chemical reactions. Strong acids are compounds that release all of their H$^+$ in solution; that is, they ionize completely. Hydrochloric acid (HCl), which is released from cells in the lining of the stomach, is a strong acid because it releases all of its H$^+$ in the stomach's watery environment. This strong acid aids in digestion and kills ingested microbes. Weak acids do not ionize completely; that is, some of their hydrogen ions remain bonded within a compound in solution. An example of a weak acid is vinegar, or acetic acid; it is called acetate after it gives up a proton.

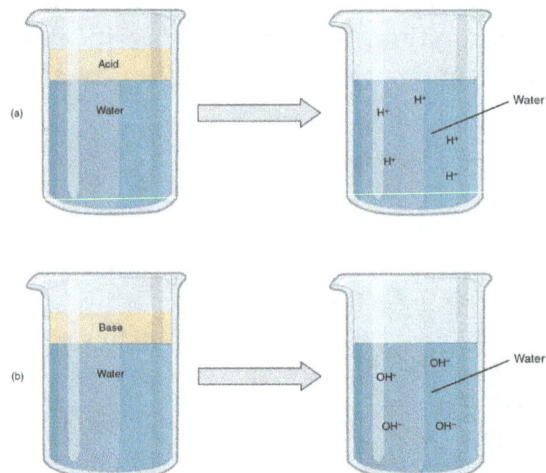

(a) In aqueous solution, an acid dissociates into hydrogen ions (H$^+$) and anions. Nearly every molecule of a strong acid dissociates, producing a high concentration of H$^+$. (b) In aqueous solution, a base dissociates into hydroxyl ions (OH$^-$) and cations. Nearly every molecule of a strong base dissociates, producing a high concentration of OH$^-$.

A base is a substance that releases hydroxyl ions (OH$^-$) in solution, or one that accepts H+ already present in solution. The hydroxyl ions (also known as hydroxide ions) or other basic substances combine with H$^+$ present to form a water molecule, thereby removing H+ and reducing the solution's acidity. Strong bases release most or all of their hydroxyl ions; weak bases release only some hydroxyl ions or absorb only a few H$^+$. Food mixed with hydrochloric acid from the stomach would burn the small intestine (the next portion of the digestive tract after the stomach), if it were not for the release of bicarbonate (HCO$_3^-$), a weak base that attracts H$^+$. Bicarbonate accepts some of the H$^+$ protons, thereby reducing the acidity of the solution.

The Concept of pH

The relative acidity or alkalinity of a solution can be indicated by its pH. A solution's pH is the negative, base-10 logarithm of the hydrogen ion (H⁺) concentration of the solution. As an example, a pH 4 solution has an H⁺ concentration that is ten times greater than that of a pH 5 solution. That is, a solution with a pH of 4 is ten times more acidic than a solution with a pH of 5. The concept of pH will begin to make more sense when you study the pH scale. The scale consists of a series of increments ranging from 0 to 14. A solution with a pH of 7 is considered neutral—neither acidic nor basic. Pure water has a pH of 7. The lower the number below 7, the more acidic the solution, or the greater the concentration of H+. The concentration of hydrogen ions at each pH value is 10 times different than the next pH. For instance, a pH value of 4 corresponds to a proton concentration of 10−4 M, or 0.0001M, while a pH value of 5 corresponds to a proton concentration of 10−5 M, or 0.00001M. The higher the number above 7, the more basic (alkaline) the solution, or the lower the concentration of H+. Human urine, for example, is ten times more acidic than pure water, and HCl is 10,000,000 times more acidic than water.

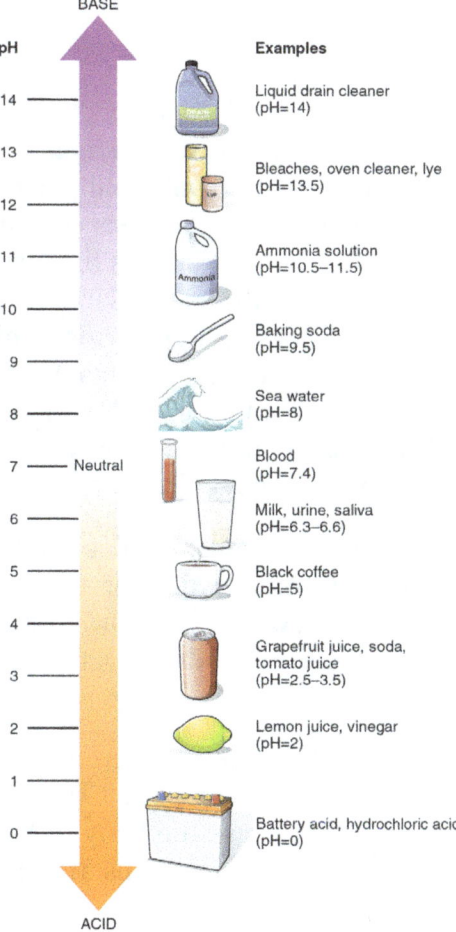

Buffers

The pH of human blood normally ranges from 7.35 to 7.45, although it is typically iden-tified as pH 7.4. At this slightly basic pH, blood can reduce the acidity resulting from the carbon dioxide (CO_2) constantly being released into the bloodstream by the trillions of cells in the body. Homeostatic mechanisms (along with exhaling CO_2 while breath-ing) normally keep the pH of blood within this narrow range. This is critical, because fluctuations—either too acidic or too alkaline—can lead to life-threatening disorders.

All cells of the body depend on homeostatic regulation of acid–base balance at a pH of approximately 7.4. The body therefore has several mechanisms for this regulation, involving breathing, the excretion of chemicals in urine, and the internal release of chemicals collectively called buffers into body fluids. A buffer is a solution of a weak acid and its conjugate base. A buffer can neutralize small amounts of acids or bases in body fluids. For example, if there is even a slight decrease below 7.35 in the pH of a bodily fluid, the buffer in the fluid—in this case, acting as a weak base—will bind the excess hydrogen ions. In contrast, if pH rises above 7.45, the buffer will act as a weak acid and contribute hydrogen ions.

Homeostatic Imbalances

The excessive acidity of acids and bases, of the blood, and other body fluids is known as acidosis. Common causes of acidosis are situations and disorders that reduce the effectiveness of breathing, especially the person's ability to exhale fully, which causes a buildup of CO_2 (and H^+) in the bloodstream. Acidosis can also be caused by metabolic problems that reduce the level or function of buffers that act as bases, or that promote the production of acids. For instance, with severe diarrhea, too much bicarbonate can be lost from the body, allowing acids to build up in body fluids. In people with poorly managed diabetes (ineffective regulation of blood sugar), acids called ketones are pro-duced as a form of body fuel. These can build up in the blood, causing a serious condi-tion called diabetic ketoacidosis. Kidney failure, liver failure, heart failure, cancer, and other disorders also can prompt metabolic acidosis.

In contrast, alkalosis is a condition in which the blood and other body fluids are too alkaline (basic). As with acidosis, respiratory disorders are a major cause; however, in respiratory alkalosis, carbon dioxide levels fall too low. Lung disease, aspirin overdose, shock, and ordinary anxiety can cause respiratory alkalosis, which reduces the normal concentration of H^+.

Metabolic alkalosis often results from prolonged, severe vomiting, which causes a loss of hydrogen and chloride ions (as components of HCl). Medications can also prompt alkalosis. These include diuretics that cause the body to lose potassium ions, as well as antacids when taken in excessive amounts, for instance by someone with persistent heartburn or an ulcer.

Aluminium Chloride

Aluminium chloride ($AlCl_3$) is a compound of aluminium and chlorine. The solid has a low melting and boiling point, and is covalently bonded. It sublimes at 178° C. Molten $AlCl_3$ conducts electricity poorly, unlike more ionic halides such as sodium chloride. It exists in the solid state as a six-coordinate layer lattice.

$AlCl_3$ adopts the " YCl_3 " structure, featuring Al^{3+} cubic close packed layered structure. In contrast, $AlBr_3$ has a more molecular structure, with the Al^{3+} centers occupying adjacent tetrahedral holes of the close-packed framework of Br^- ions. Upon melting $AlCl_3$ gives the dimer Al_2Cl_6, which can vaporize. At higher temperatures this Al_2Cl_6 dimer dissociates into trigonal planar $AlCl_3$, which is structurally analogous to BF_3.

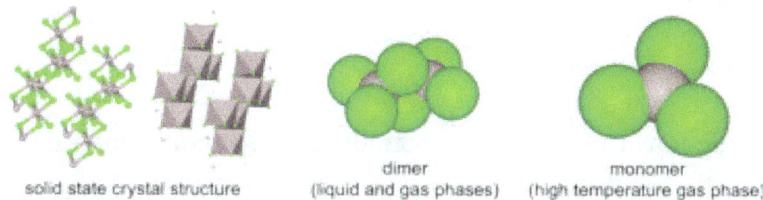

solid state crystal structure   dimer   monomer
                          (liquid and gas phases)   (high temperature gas phase)

Aluminium chloride is highly deliquescent, and it can explode in contact with water because of the high heat of hydration. It partially hydrolyses with H_2O, forming some hydrogen chloride and/or hydrochloric acid. Aqueous solutions of $AlCl_3$ are fully ionized, and thus conduct electricity well. Such solutions are found to be acidic, indicating that partial hydrolysis of the Al^{3+} ion is occurring. This can be described (simplified) as:

$$[Al(H_2O)_6]^{3+} \xrightleftharpoons{H_2O} [Al(OH)(H_2O)_5]^{2+}$$
$$+ H_3O^+$$

$AlCl_3$ is probably the most commonly used non-Bronsted Lewis acid and also one of the most powerful. It finds widespread application in the chemical industry as a catalyst for Friedel-Crafts reactions, both acylations and alkylations. It also finds use in polymerization and isomerization reactions of hydrocarbons. Aluminium chloride, like similar compounds such as Aluminium chlorohydrate, is also commonly used as an antiperspirant.

Aluminium also forms a lower chloride, aluminium(I) chloride (AlCl), but this is very unstable and only known in the vapour phase.

Chemical Properties

Aluminium chloride is a powerful Lewis acid, capable of forming stable Lewis acid-base adducts with even weak Lewis bases such as benzophenone or mesitylene. Not surprisingly it forms $AlCl_4^-$ in the presence of chloride ion.

In water, partial hydrolysis forms HCl gas or H_3O^+, as described in the overview above. Aqueous solutions behave similarly to other aluminium salts containing hydrated Al^{3+} ions - for example giving a gelatinous precipitate of aluminium hydroxide upon reaction with the correct quantity of aqueous sodium hydroxide:

$$AlCl_3\,(aq) + 3\,NaOH(aq) \longrightarrow Al(OH)_3\,(s) + 3\,NaCl(aq)$$

Synthesis

Aluminium chloride is manufactured on a large scale by the exothermic reaction of aluminium metal with chlorine or hydrogen chloride at temperatures between 650 to $750\,°C\,(1,202\,to\,1,382\,°F)$.

$$2\,Al + 3\,Cl_2 \longrightarrow 2\,AlCl_3$$

$$2\,Al + 6\,HCl \longrightarrow 2\,AlCl_3 + 3\,H_2$$

Aluminum chloride may be formed via a single displacement reaction between copper chloride and aluminum metal.

$$2\,Al + 3\,CuCl_2 \longrightarrow 2\,AlCl_3 + 3\,Cu$$

In the US in 1993, approximately 21,000 tons were produced, not counting the amounts consumed in the production of aluminium.

Hydrated aluminium trichloride is prepared by dissolving aluminium oxides in hydrochloric acid. Metallic aluminum also readily dissolves in hydrochloric acid — releasing hydrogen gas and generating considerable heat. Heating this solid does not produce anhydrous aluminium trichloride, the hexahydrate decomposes to aluminium hydroxide when heated:

$$Al(H_2O)_6\,Cl_3 \longleftarrow Al(OH)_3 + 3\,HCl + 3\,H_2O$$

Aluminium also forms a lower chloride, aluminium(I) chloride (AlCl), but this is very unstable and only known in the vapour phase.

Uses

Anhydrous Aluminium Trichloride

$AlCl_3$ is probably the most commonly used Lewis acid and also one of the most powerful. It finds application in the chemical industry as a catalyst for Friedel–Crafts reactions, both acylations and alkylations. Important products are detergents and ethylbenzene. It also finds use in polymerization and isomerization reactions of hydrocarbons.

The Friedel–Crafts reaction is the major use for aluminium chloride, for example in the preparation of anthraquinone (for the dyestuffs industry) from benzene and phosgene. In the general Friedel–Crafts reaction, an acyl chloride or alkyl halide reacts with an aromatic system as shown:

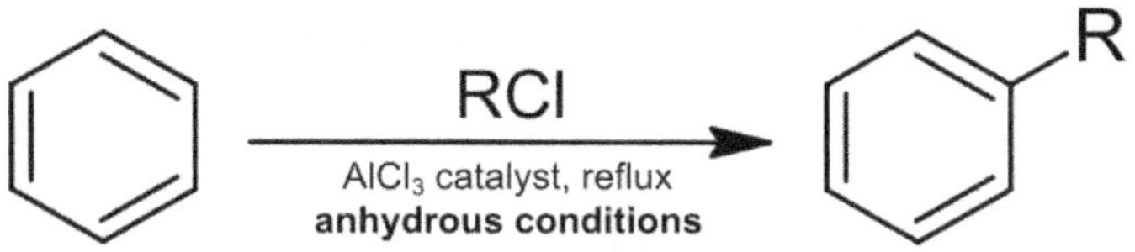

The alkylation reaction is more widely used than the acylation reaction, although its practice is more technically demanding because the reaction is more sluggish. For both reactions, the aluminium chloride, as well as other materials and the equipment, should be dry, although a trace of moisture is necessary for the reaction to proceed. A general problem with the Friedel–Crafts reaction is that the aluminium chloride catalyst sometimes is required in full stoichiometric quantities, because it complexes strongly with the products. This complication sometimes generates a large amount of corrosive waste. For these and similar reasons, more recyclable or environmentally benign catalysts have been sought. Thus, the use of aluminium chloride in some applications is being displaced by zeolites.

Aluminium chloride can also be used to introduce aldehyde groups onto aromatic rings, for example via the Gattermann-Koch reaction which uses carbon monoxide, hydrogen chloride and a copper(I) chloride co-catalyst.

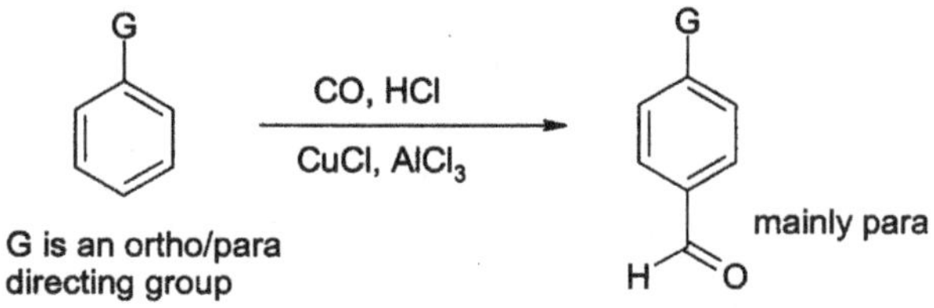

Aluminium chloride finds a wide variety of other applications in organic chemistry. For example, it can catalyse the "ene reaction", such as the addition of 3-buten-2-one (methyl vinyl ketone) to carvone:

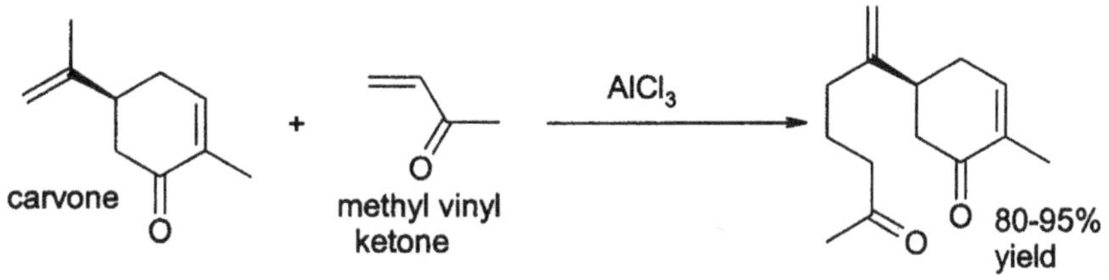

$AlCl_3$ is also widely used for polymerization and isomerization reactions of hydrocarbons. Important examples include the manufacture of ethylbenzene, which used to make styrene and thus polystyrene, and also production of dodecylbenzene, which is used for making detergents.

Aluminium chloride combined with aluminium in the presence of an arene can be used to synthesize bis(arene) metal complexes, e.g. bis(benzene)chromium, from certain metal halides via the so-called Fischer-Hafner synthesis.

Hydrated Aluminium Chlorides

The dihydrate has few applications, but aluminium chlorohydrate is a common component in antiperspirants at low concentrations. Hyperhidrosis sufferers need a much higher concentration (12% or higher), sold under such brand names as Xeransis, Drysol, DryDerm, sunsola, Maxim, Odaban, CertainDri, B+Drier, Chlorhydrol, Anhydrol Forte and Driclor.

Industrial Uses

The primary uses of aluminum chloride are in manufacturing and industry. First and foremost, it's a component in the production of aluminum, in metallurgy, and as an ingredient in aluminum smelting. It's also used in manufacturing petrochemicals like ethylbenzene and alkylbenzene. Certain kinds of pharmaceuticals require aluminum chloride as an ingredient. Its many other applications include the production of paint, synthetic rubber, lubricants, wood preservatives, and some organic chemicals. This is a versatile compound.

In the Home

A use that may come closer to home for most people is that there's a tiny bit of aluminum chloride in many antiperspirants. In fact, larger amounts of it are in prescription antiperspirants. Aluminum chloride works in deodorant by combining with electrolytes in the skin to create a gel plug in the sweat glands. It also has a slightly astringent effect on the pores, causing them to contract, which keeps the pores from releasing sweat. Because of this ability to contract pores, it's also an ingredient in cosmetic astringents.

Dangers of Using It

Aluminum chloride is potentially hazardous to humans, particularly in its anhydrous form. It's highly corrosive and can lead to serious damage if it's inhaled, ingested, or touches the skin. Inhalation can lead to severe irritation of the nose and throat, difficulty in breathing, headache, nausea, and vomiting. Severe eye irritation and the risk of permanent damage to the eyes can happen if it gets in the eyes. Skin contact can lead to extreme irritation or severe burns. Ingestion can result in permanent damage to the digestive system.

Preventing Exposure

Household forms of aluminum chloride are not harmful to humans. Exposure to aluminum chloride generally only happens in a laboratory or industrial setting. In these cases, the managers of these facilities must take precautions to make sure people aren't exposed. These include respiratory protection and suitable ventilation, chemical resistant PVC gloves, chemical-proof goggles or face shields, and PVC aprons and boots when splashing is a risk. Management should offer decontamination showers and eyewash stations as a precaution.

Aluminum chloride is a compound of aluminum and chloride that's widely used in petroleum refining and the manufacturing of many products. In addition, antiperspirants and cosmetic astringents use this compound.

Ammonia

Ammonia (NH_3) is a colourless, pungent gas composed of nitrogen and hydrogen. It is the simplest stable compound of these elements and serves as a starting material for the production of many commercially important nitrogen compounds.

Structure

Ammonia is a covalent atom. It is seen as a dot structure. The particle is shaped because of the overlap of orbitals of three hydrogen atoms and three sp3 hybrid orbitals of nitrogen in the structure as the central atom. The fourth sp^3 hybrid orbital is involved by a lone pair.

This provides a trigonal pyramidal shape to the compound. The H-N-H bond edge is 107.3°, which is somewhat not exactly the tetrahedral edge of 109°28. This is on the grounds that the bond pair-lone pair repulsions push the N-H bonds somewhat inwards. In solid and liquid states, ammonia is related through hydrogen bonds.

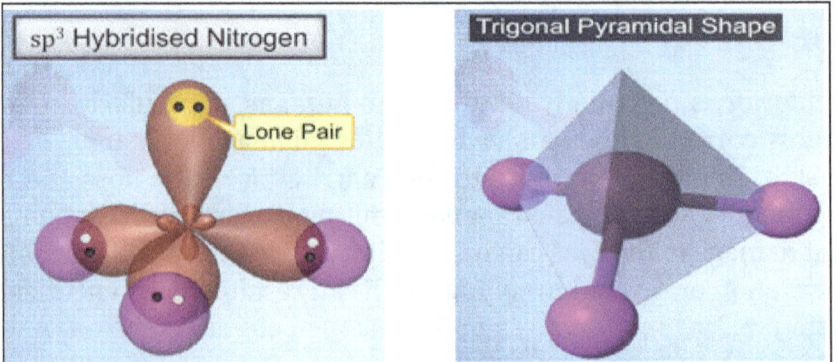

Manufacture of Ammonia

The manufacture of ammonia from nitrogen and hydrogen takes place in two main stages:

a) The manufacture of hydrogen.

b) The synthesis of ammonia (the Haber Process).

The manufacture of hydrogen involves several distinct processes. Figure shows their sequence and the location within an ammonia plant (steps1-5). The converter used to make ammonia from the hydrogen is also shown (step 6). What occurs in each of these steps is described below.

Western Australia:

1) Desulfurisation units

2) Primary reformer

3) High temperature and low temperature shift reactors

4) Carbon dioxide absorber

5) Carbon dioxide stripper (recovery of the pure solvent, ethanolamine)

6) Ammonia converter

7) Ammonia storage as liquid

8) Pipeline to the ship for export

a) The manufacture of hydrogen

Hydrogen is produced from a variety of feedstocks, mostly from natural gas, coal or naphtha. The ways in which hydrogen is obtained from these feedstocks are dealt with separately.

Hydrogen from Natural Gas (Methane)

This involves two stages:

i) The manufacture of synthesis gas (a mixture of carbon monoxide and hydrogen (steam reforming)).

ii) The removal of the carbon monoxide and production of a mixture of hydrogen and nitrogen (the shift reaction).

The Manufacture of Synthesis Gas

Whichever way the methane is obtained, it will contain some organic sulfur compounds and hydrogen sulfide, both of which must be removed. Otherwise, they will poison the catalyst needed in the manufacture of synthesis gas. In the desulfurisation unit, the organic sulfur compounds are often first converted into hydrogen sulfide, prior to reaction with zinc oxide. The feedstock is mixed with hydrogen and passed over a catalyst of mixed oxides of cobalt and molybdenum on an inert support (a specially treated alumina) at ca 700 K.

$$R\text{-}SH(g) \longrightarrow R\text{-}H(g) + H_2S(g)$$

Then the gases are passed over zinc oxide at ca 700 K and hydrogen sulfide is removed:

$$ZnO(s) + H_2S(g) \longrightarrow ZnS(s) + H_2O(g)$$

Primary steam reforming converts methane and steam to synthesis gas, a mixture of carbon monoxide and hydrogen

$$CH_4(g) + H_2O(g) \rightleftharpoons CO(g) + 3H_2(g) \quad \Delta H^{\ominus} = +210\,kj\,mol^{-1}$$

High temperatures and low pressures favour the formation of the products (Le Chatelier's Principle). In practice, the reactants are passed over a catalyst of nickel, finely divided on the surface of a calcium oxide/aluminium oxide support contained in vertical nickel alloy tubes. The tubes, up to 350 in parallel, are heated in a furnace above 1000 K and under a pressure of ca 30 atm. This is an example of a tubular reactor.

Secondary steam reforming reacts oxygen from the air with some of the hydrogen present and the resulting mixture is passed over a nickel catalyst. The steam and heat produced from the combustion reforms most of the residual methane. Among the key reactions are:

$$2H_2(g) + O_2(g) \longrightarrow 2H_2O(g) \qquad \Delta H^{\ominus} = -482\,KJ\,mol^{-1}$$
$$CH_4(g) + H_2O(g) \rightleftharpoons CO(g) + 3H_2(g) \qquad \Delta H^{\ominus} = -482\,KJ\,mol^{-1}$$

The emerging gas from this net exothermic stage is at ca 1200 K and is cooled in heat exchangers. The steam formed from the water used in cooling the gases is used to operate turbines and thus compressors and to preheat reactants.

Some recent designs use waste heat from the secondary reformer directly to provide heat for the primary reformer.

At this stage the gas contains hydrogen, nitrogen, carbon monoxide and carbon dioxide and about 0.25% methane. As air contains 1% argon, this also accumulates in the synthesis gas.

The Shift Reaction

This process converts carbon monoxide to carbon dioxide, while generating more hydrogen.

It takes place in two stages. In the first, the high temperature shift reaction, the gas is mixed with steam and passed over an iron/chromium(III) oxide catalyst at ca 700 K in a fixed bed reactor. This decreases the carbon monoxide concentration from 11%:

$$CO(g) + H_2O(g) \rightleftharpoons CO_2(g) + H_2(g) \quad \Delta H^\ominus = -42\,KJ\,mol^{-1}$$

In the second stage, the low temperature shift reaction, the mixture of gases is passed over a copper-zinc catalyst at ca 500 K. The carbon monoxide concentration is further reduced to 0.2%.

The reaction is done in two stages for several reasons. The reaction is exothermic. However, at high temperature, the exit concentration of carbon monoxide is still quite high, due to equilibrium control. The copper catalyst used in the low temperature stage is very sensitive to high temperatures, and could not operate effectively in the high temperature stage. Thus, the bulk of the reaction is carried out at high temperature to recover most of the heat. The gas is then removed at low temperature, where the equilibrium is much more favorable, on the very active but unstable copper catalyst.

The gas mixture now contains about 18% carbon dioxide which is removed by scrubbing the gas with a solution of a base, using one of several available methods. One that is favoured is an organic base (in the carbon dioxide absorber), a solution of an ethanolamine, often 2,2'-(methylimino)bis-ethanol (N-methyl diethanolamine).

The carbon dioxide is released on heating the solution in the carbon dioxide stripper). Much of it is liquefied and sold, for example, for carbonated drinks, as a coolant for nuclear power stations and for promoting the growth of plants in greenhouses.

The last traces of oxides of carbon are removed by passing the gases over a nickel catalyst at 600 K:

$$CO(g) + 3H(g) \rightleftharpoons CH4(g) + H_2O(g)$$

$$CO_2(g) + 4H_2(g) \rightleftharpoons CH_4(g) + 2H_2O(g)$$

This process is known as methanation. A gas is obtained of typical composition: 74% hydrogen, 25% nitrogen, 1% methane, together with some argon.

Hydrogen from Naphtha

If naphtha is used as the feedstock, an extra reforming stage is needed. The naphtha is heated to form a vapour, mixed with steam and passed through tubes, heated at 750 K and packed with a catalyst, nickel supported on a mixture of aluminium and magnesium oxides. The main product is methane together with oxides of carbon, and is then processed by steam reforming, as if it was natural gas, followed by the shift reaction.

Hydrogen from Coal

If coal is used, it is first finely ground and heated in an atmosphere of oxygen and steam. Some of the coal burns very rapidly in oxygen (in less than 0.1 s) causing the temperature in the furnace to rise and the rest of the coal reacts with the steam:

$$C(s) + H_2O(g) \longrightarrow CO(g) + H_2(g)$$

The gas emitted contains ca 55% carbon monoxide, 30% hydrogen, 10% carbon dioxide and small amounts of methane and other hydrocarbons. This mixture is treated by the shift reaction.

The main problems of using coal includes the large amounts of sulfur dioxide and trioxide generated in burning coal and the significant amounts of other impurities such as arsenic and bromine, all of which are very harmful to the atmosphere and all of which are severe poisons to the catalysts in the process. There is also a massive problem with disposal of the ash.

Hydrogen from Biomass

Synthesis gas gas can be produced from biopmass. The process is outlined in the unit on biorefineries.

(a) The manufacture of ammonia (The Haber Process)

$$N_2(g) + 3H(g) \rightleftharpoons 2NH_3(g) \quad \Delta H^{\oplus} = -92\,KJ\,mol^{-1}$$

The heart of the process is the reaction between hydrogen and nitrogen in a fixed bed reactor. The gases, in stoichiometric proportions, are heated and passed under pressure over a catalyst.

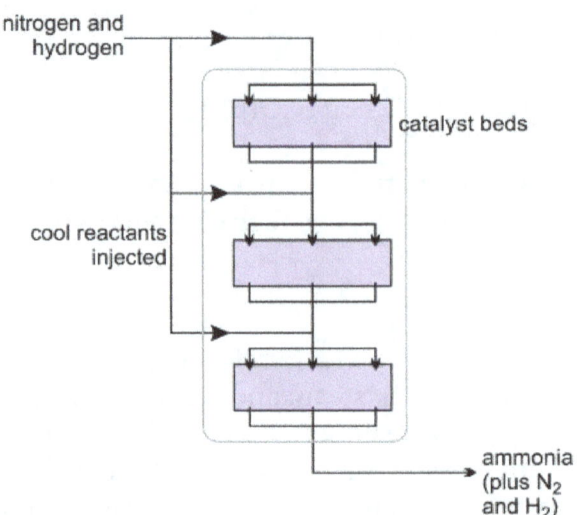

Diagram illustrating a conventional synthesis reactor (a converter).

The proportion of ammonia in the equilibrium mixture increases with increasing pressure and with falling temperature (Le Chatelier's Principle). Quantitative data are given in Table below. To obtain a reasonable yield and favourable rate, high pressures, moderate temperatures and a catalyst are used.

Pressure/atm	Percentage ammonia present at equilibrium at a range of temperatures					
	373 K	473 K	573 K	673 K	773 K	973 K
10	-	50.7	14.7	3.9	1.2	0.2
25	91.7	63.6	27.4	8.7	2.9	-
50	94.5	74.0	39.5	15.3	5.6	1.1
100	96.7	81.7	52.5	25.2	10.6	2.2
200	98.4	89.0	66.7	38.8	18.3	-
400	99.4	94.6	79.7	55.4	31.9	-
1000	-	98.3	92.6	79.8	57.5	12.9

Percentage, by volume, of ammonia in the equilibrium mixture for the reaction between nitrogen and hydrogen at a range of temperatures and pressures.

A wide range of conditions are used, depending on the construction of the reactor. Temperatures used vary between 600 and 700 K, and pressures between 100 and 200 atmospheres. Much work is being done to improve the effectiveness of the catalyst so that pressures as low as 50 atmospheres can be used.

As the reaction is exothermic, cool reactants (nitrogen and hydrogen) are added to reduce the temperature of the reactors.

The ammonia is usually stored on site (step 7) and pumped to another part of the plant where it is converted into a fertilizer (urea or an ammonium salt). However it is sometimes transported by sea or by road, to be used in another plant.

In a plant in Western Australia, the ammonia is transferred by pipeline to a nearby harbour and transported by ship. This one is carrying about 40 000 tonnes of liquefied ammonia.

The original catalyst that Haber used was Fe_3O_4, which was reduced by the reactant, hydrogen, to iron. Much work was done to improve the catalyst and it was found that a small amount of potassium hydroxide was effective as a promoter.

Physical Properties Of Ammonia

Ammonia is a colorless gas with a sharp, penetrating odour. Its boiling point is −33.35 °C (−28.03 °F), and its freezing point is −77.7 °C (−107.8 °F). It has a high heat of vaporization (23.3 kilojoules per mole at its boiling point) and can be handled as a liquid in thermally insulated containers in the laboratory. (The heat of vaporization of a substance is the number of kilojoules needed to vaporize one mole of the substance with no change in temperature.) The ammonia molecule has a trigonal pyramidal shape with the three hydrogen atoms and an unshared pair of electrons attached to the nitrogen atom. It is a polar molecule and is highly associated because of strong intermolecular hydrogen bonding. The dielectric constant of ammonia (22 at −34 °C [−29 °F]) is lower than that of water (81 at 25 °C [77 °F]), so it is a better solvent for organic materials. However, it is still high enough to allow ammonia to act as a moderately good ionizing solvent. Ammonia also self-ionizes, although less so than does water.

$$2\,NH_3 \rightleftharpoons NH_4^+ + NH_2^-$$

Chemical Reactivity Of Ammonia

The combustion of ammonia proceeds with difficulty but yields nitrogen gas and water.

$$4\,NH_3 + 3\,O_2 + heat \longrightarrow 2\,N_2 + 6\,H_2O$$

However, with the use of a catalyst and under the correct conditions of temperature, ammonia reacts with oxygen to produce nitric oxide, NO, which is oxidized to nitrogen dioxide, NO_2, and is used in the industrial synthesis of nitric acid.

Ammonia readily dissolves in water with the liberation of heat.

$$NH_3 + H_2O \rightleftharpoons NH_4^+ + OH^-$$

These aqueous solutions of ammonia are basic and are sometimes called solutions of ammonium hydroxide ($NH\ OH$). The equilibrium, however, is such that a 1.0-molar solution of NH_3 provides only 4.2 millimoles of hydroxide ion. The hydrates $NH_3 \cdot H_2O, 2\,NH_3 \cdot H_2O$, and $NH_3 \cdot 2\,H_2O$ exist and have been shown to consist of ammonia and water molecules linked by intermolecular hydrogen bonds.

Liquid ammonia is used extensively as a nonaqueous solvent. The alkali metals as well as the heavier alkaline-earth metals and even some inner transition metals dissolve in liquid ammonia, producing blue solutions. Physical measurements, including electrical-conductivity studies, provide evidence that this blue colour and electrical current are due to the solvated electron.

Liquid ammonia is used extensively as a nonaqueous solvent. The alkali metals as well as the heavier alkaline-earth metals and even some inner transition metals dissolve in liquid ammonia, producing blue solutions. Physical measurements, including electrical-conductivity studies, provide evidence that this blue colour and electrical current are due to the solvated electron.

$$metal\,(dispersed) \rightleftharpoons metal(NH_3)_x \rightleftharpoons M^+(NH_3)_x + e^-(NH_3)_y$$

These solutions are excellent sources of electrons for reducing other chemical species. As the concentration of dissolved metal increases, the solution becomes a deeper blue in colour and finally changes to a copper-coloured solution with a metallic lustre. The electrical conductivity decreases, and there is evidence that the solvated electrons associate to form electron pairs.

$$2\,e^-(NH_3)_y \rightleftharpoons e_2(NH_3)_y$$

Most ammonium salts also readily dissolve in liquid ammonia.

Derivatives of Ammonia

Two of the more important derivatives of ammonia are hydrazine and hydroxylamine.

Hydrazine

Hydrazine N_2H_4, is a molecule in which one hydrogen atom in NH_3 is replaced by an $-NH_2$ group. The pure compound is a colourless liquid that fumes with a slight odour similar to that of ammonia. In many respects it resembles water in its physical properties. It has a melting point of 2 °C (35.6 °F), a boiling point of 113.5 °C (236.3 °F), a high dielectric constant (51.7 at 25 °C [77 °F]), and a density of 1 gram per cubic cm. As with water and ammonia, the principal intermolecular force is hydrogen bonding.

Hydrazine is best prepared by the Raschig process, which involves the reaction of an aqueous alkaline ammonia solution with sodium hypochlorite (NaOCl).

$$2\,NH_3 + NaOCl \longrightarrow N_2H_4 + NaCl + H_2O$$

This reaction is known to occur in two main steps. Ammonia reacts rapidly and quantitatively with the hypochlorite ion OCl^-, to produce chloramine, NH_2Cl, which reacts further with more ammonia and base to produce hydrazine.

$$NH_3 + OCl^- \longrightarrow NH_2Cl + OH^-$$

$$NH_2Cl + NH_3 + NaOH \longrightarrow N_2H_4 + NaCl + H_2O$$

In this process there is a detrimental reaction that occurs between hydrazine and chloramine and that appears to be catalyzed by heavy metal ions such as Cu^{2+}. Gelatin is added to this process to scavenge these metal ions and suppress the side reaction.

$$N_2H_4 + 2\,NH_2Cl \longrightarrow 2\,NH_4Cl + N_2$$

When hydrazine is added to water, two different hydrazinium salts are obtained. $N_2H_5^+$ salts can be isolated, but $N_2H_6^{2+}$ salts are normally extensively hydrolyzed.

$$N_2H_4 + H_2O \rightleftharpoons N_2H_5^+ + OH^-$$

$$N_2H_5^+ + H_2O \rightleftharpoons N_2H_6^{2+} + OH^-$$

Hydrazine burns in oxygen to produce nitrogen gas and water, with the liberation of a substantial amount of energy in the form of heat.

$$N_2H_4 + O_2 \longrightarrow N_2 + 2\,H_2O + heat$$

As a result, the major noncommercial use of this compound (and its methyl derivatives) is as a rocket fuel. Hydrazine and its derivatives have been used as fuels in guided missiles, spacecraft (including the space shuttles), and space launchers. For example, the Apollo program's Lunar Module was decelerated for landing, and launched from the Moon, by the oxidation of a 1:1 mixture of methyl hydrazine, H_3CNHNH_2, , and 1,1-dimethylhydrazine, $(H_3C)_2 NNH_2$, with liquid dinitrogen tetroxide, N_2O_4. Three tons of the methyl hydrazine mixtures were required for the landing on the Moon, and about one ton was required for the launch from the lunar surface. The major commercial uses of hydrazine are as a blowing agent (to make holes in foam rubber), as a reducing agent, in the synthesis of agricultural and medicinal chemicals, as algicides, fungicides, and insecticides, and as plant growth regulators.

Hydroxylamine

Hydroxylamine, NH_2OH, may be thought of as being derived from ammonia by replacement of a hydrogen atom with a hydroxyl group (–OH). The pure compound is a colourless solid that is hygroscopic (rapidly absorbs water) and thermally unstable. It must be stored at 0 °C (32 °F) so that it will not decompose. It melts at 33 °C (91.4 °F), has a density of 1.2 grams per cubic cm at 33 °C, and has a high dielectric constant (ε = 78). Aqueous solutions of hydroxylamine are not as strongly basic as either ammonia or hydrazine. Hydroxylamine can be prepared by a number of reactions. A laboratory synthesis involves the reduction of aqueous potassium nitrite, KNO_2, or nitrous acid, HNO_2, with the hydrogen sulfite ion,. In HSO_3^- general, hydroxylamine is stored and used as an aqueous solution or as a salt (for example, $NH_3OH^+NO_3^-$ It is often used in the preparation of oximes.

Uses

Fertilizer

Globally, approximately 88% (as of 2014) of ammonia is used as fertilizers either as its salts, solutions or anhydrously. When applied to soil, it helps provide increased yields of crops such as maize and wheat. 30% of agricultural nitrogen applied in the USA is in the form of anhydrous ammonia and worldwide 110 million tonnes are applied each year.

Precursor to Nitrogenous Compounds

Ammonia is directly or indirectly the precursor to most nitrogen-containing compounds. Virtually all synthetic nitrogen compounds are derived from ammonia. An important derivative is nitric acid. This key material is generated via the Ostwald process by oxidation of ammonia with air over a platinum catalyst at 700–850 °C (1,292–1,562 °F), ~9 atm. Nitric oxide is an intermediate in this conversion:

$$NH_3 + 2 O_2 \rightarrow HNO_3 + H_2O$$

Nitric acid is used for the production of fertilizers, explosives, and many organonitrogen compounds.

Ammonia is also used to make the following compounds:

- Hydrazine, in the Olin Raschig process and the peroxide process
- Hydrogen cyanide, in the BMA process and the Andrussow process
- Hydroxylamine and ammonium carbonate, in the Raschig process
- Phenol, in the Raschig–Hooker process
- Urea, in the Bosch–Meiser urea process and in Wöhler synthesis
- Amino acids, using Strecker amino-acid synthesis
- Acrylonitrile, in the Sohio process

Ammonia can also be used to make compounds in reactions which are not specifically named. Examples of such compounds include: ammonium perchlorate, ammonium nitrate, formamide, dinitrogen tetroxide, alprazolam, ethanolamine, ethyl carbamate, hexamethylenetetramine, and ammonium bicarbonate.

As a Cleaner

Household ammonia is a solution of NH_3 in water (i.e., ammonium hydroxide) used as a general purpose cleaner for many surfaces. Because ammonia results in a relatively streak-free shine, one of its most common uses is to clean glass, porcelain and stainless steel. It is also frequently used for cleaning ovens and soaking items to loosen baked-on grime. Household ammonia ranges in concentration by weight from 5 to 10% ammonia. United States manufacturers of cleaning products are required to provide the product's material safety data sheet which lists the concentration used.

Fermentation

Solutions of ammonia ranging from 16% to 25% are used in the fermentation industry as a source of nitrogen for microorganisms and to adjust pH during fermentation.

Antimicrobial Agent for Food Products

As early as in 1895, it was known that ammonia was "strongly antiseptic, it requires 1.4 grams per litre to preserve beef tea." In one study, anhydrous ammonia destroyed 99.999% of zoonotic bacteria in 3 types of animal feed, but not silage. Anhydrous ammonia is currently used commercially to reduce or eliminate microbial contamination of beef.Lean finely textured beef in the beef industry is made from fatty beef trimmings (c. 50–70% fat) by removing the fat using heat and centrifugation, then treating it with ammonia to kill *E. coli*. The process was deemed effective and safe by the US Department of Agriculture based on a study that found that the treatment reduces *E. coli* to unde-tectable levels. There have been safety concerns about the process as well as consumer complaints about the taste and smell of beef treated at optimal levels of ammonia. The level of ammonia in any final product has not come close to toxic levels to humans.

Minor and Emerging Uses

Refrigeration – R717

Because of ammonia's vaporization properties, it is a useful refrigerant. It was com-monly used before the popularisation of chlorofluorocarbons (Freons). Anhydrous am-monia is widely used in industrial refrigeration applications and hockey rinks because of its high energy efficiency and low cost. It suffers from the disadvantage of toxici-ty, which restricts its domestic and small-scale use. Along with its use in modern va-por-compression refrigeration it is used in a mixture along with hydrogen and water in absorption refrigerators. The Kalina cycle, which is of growing importance to geother-mal power plants, depends on the wide boiling range of the ammonia–water mixture. Ammonia coolant is also used in the S1 radiator aboard the International Space Station in two loops which are used to regulate the internal temperature and enable tempera-ture dependent experiments.

For remediation of Gaseous Emissions

Ammonia is used to scrub SO_2 from the burning of fossil fuels, and the resulting product is converted to ammonium sulfate for use as fertilizer. Ammonia neutral-izes the nitrogen oxides (NO_x) pollutants emitted by diesel engines. This technol-ogy, called SCR (selective catalytic reduction), relies on a vanadia-based catalyst. Ammonia may be used to mitigate gaseous spills of phosgene.

As a Fuel

Ammoniacal Gas Engine Streetcar in New Orleans drawn by Alfred Waud in 1871.

The X-15 aircraft used ammonia as one component fuel of its rocket engine

The raw energy density of liquid ammonia is 11.5 MJ/L, which is about a third that of diesel. Although it can be used as a fuel, for a number of reasons this has never been common or widespread. In addition to direct utilization of ammonia as a fuel in combustion engines, there is also the opportunity to convert ammonia back to hydrogen, where it can be used to power hydrogen fuel cells or directly within high-temperature fuel cells.

Ammonia engines or ammonia motors, using ammonia as a working fluid, have been proposed and occasionally used. The principle is similar to that used in a fireless locomotive, but with ammonia as the working fluid, instead of steam or compressed air. Ammonia engines were used experimentally in the 19th century by Goldsworthy Gurney in the UK and the St. Charles Avenue Streetcar line in New Orleans in the 1870s and 1880s, and during World War II ammonia was used to power buses in Belgium.

Ammonia is sometimes proposed as a practical alternative to fossil fuel for internal combustion engines. Its high octane rating of 120 and low flame temperature allows the use of high compression ratios without a penalty of high NOx production. Since ammonia contains no carbon, its combustion cannot produce carbon monoxide, hydrocarbons or soot.

However ammonia cannot be easily used in existing Otto cycle engines because of its very narrow flammability range, and there are also other barriers to widespread automobile usage. In terms of raw ammonia supplies, plants would have to be built to increase production levels, requiring significant capital and energy sources. Although it is the second most produced chemical, the scale of ammonia production is a small fraction of world petroleum usage. It could be manufactured from renewable energy sources, as well as coal or nuclear power. The 60 MW Rjukan dam in Telemark, Norway produced ammonia for many years from 1913, providing fertilizer for much of Europe.

Despite this, several tests have been done. In 1981, a Canadian company converted a 1981 Chevrolet Impala to operate using ammonia as fuel. In 2007, a University of Michigan pickup powered by ammonia drove from Detroit to San Francisco as part of a demonstration, requiring only one fill-up in Wyoming.

Compared to hydrogen as a fuel, ammonia is much more energy efficient, and hydrogen could be produced, stored, and delivered at a much lower cost as ammonia rather than as compressed and/or cryogenic hydrogen. The conversion of ammonia to hydrogen via the sodium-amide process, either as a catalyst for combustion or as fuel for a proton exchange membrane fuel cell, is another possibility. Conversion to hydrogen would allow the storage of hydrogen at nearly 18 wt% compared to ~5% for gaseous hydrogen under pressure.

Rocket engines have also been fueled by ammonia. The Reaction Motors XLR99 rocket engine that powered the X-15 hypersonic research aircraft used liquid ammonia. Although not as powerful as other fuels, it left no soot in the reusable rocket engine, and its density approximately matches the density of the oxidizer, liquid oxygen, which simplified the aircraft's design.

As a Stimulant

Anti-meth sign on tank of anhydrous ammonia, Otley, Iowa. Anhydrous ammonia is a common farm fertilizer that is also a critical ingredient in making methamphetamine. In 2005, Iowa used grant money to give out thousands of locks to prevent criminals from getting into the tanks.

Ammonia, as the vapor released by smelling salts, has found significant use as a respiratory stimulant. Ammonia is commonly used in the illegal manufacture of methamphetamine through a Birch reduction. The Birch method of making methamphetamine is dangerous because the alkali metal and liquid ammonia are both extremely reactive, and the temperature of liquid ammonia makes it susceptible to explosive boiling when reactants are added.

Textile

Liquid ammonia is used for treatment of cotton materials, giving properties like mercerisation, using alkalis. In particular, it is used for prewashing of wool.

Lifting Gas

At standard temperature and pressure, ammonia is less dense than atmosphere and has approximately 45-48% of the lifting power of hydrogen or helium. Ammonia has sometimes been used to fill weather balloons as a lifting gas. Because of its relatively high boiling point (compared to helium and hydrogen), ammonia could potentially be refrigerated and liquefied aboard an airship to reduce lift and add ballast (and returned to a gas to add lift and reduce ballast).

Woodworking

Ammonia has been used to darken quartersawn white oak in Arts & Crafts and Mission-style furniture. Ammonia fumes react with the natural tannins in the wood and cause it to change colours.

Energy Carrier

Ammonia can be manufactured from solar energy, air and water. This is an efficient way to package hydrogen into a chemical that is much cheaper to store and transport than pure hydrogen be it as gas or as liquid. In fact, per volume ammonia holds more hydrogen than does liquid hydrogen. Ammonia may be the key to overcome not only the daily but also the seasonal fluctuations of renewable energy sources.

This approach will solve many of the problems foreseen for the proposed Hydrogen economy, that instead could be replaced by an Ammonia economy, essentially still a hydrogen economy.

In early August 2018, scientists from Australia's Commonwealth Scientific and Industrial Research Organisation (CSIRO) announced the success of developing a process to release hydrogen from ammonia and harvest that at ultra-high purity as a fuel for cars. This uses a special membrane. Two demonstration fuel cell vehicles have the technology, a Hyundai Nexo and Toyota Mirai.

Safety Precautions

The world's longest ammonia pipeline, running from the TogliattiAzot plant in Russia to Odessa in Ukraine.

The U. S. Occupational Safety and Health Administration (OSHA) has set a 15-minute exposure limit for gaseous ammonia of 35 ppm by volume in the environmental air and an 8-hour exposure limit of 25 ppm by volume. NIOSH recently reduced the IDLH from 500 to 300 based on recent more conservative interpretations of original research in 1943. IDLH (Immediately Dangerous to Life and Health) is the level to which a healthy worker can be exposed for 30 minutes without suffering irreversible health effects. Other organizations have varying exposure levels. U.S. Navy Standards [U.S. Bureau of Ships 1962] maximum allowable concentrations (MACs):continuous exposure (60 days): 25 ppm / 1 hour: 400 ppm Ammonia vapour has a sharp, irritating, pungent odour that acts as a warning of potentially dangerous exposure. The average odour threshold is 5 ppm, well below any danger or damage. Exposure to very high concentrations of gaseous ammonia can result in lung damage and death. Although ammonia is regulated in the United States as a non-flammable gas, it still meets the definition of a material that is toxic by inhalation and requires a hazardous safety permit when transported in quantities greater than 13,248 L (3,500 gallons). Household products containing ammonia (i.e., Windex) should never be used in conjunction with products containing bleach, as the resulting chemical reaction produces highly toxic fumes.

Liquid ammonia is dangerous because it is hygroscopic and because it can freeze flesh.

Toxicity

The toxicity of ammonia solutions does not usually cause problems for humans and other mammals, as a specific mechanism exists to prevent its build-up in the bloodstream. Ammonia is converted to carbamoyl phosphate by the enzyme carbamoyl phosphate synthetase, and then enters the urea cycle to be either incorporated into amino acids or excreted in the urine. Fish and amphibians lack this mechanism, as they can usually eliminate ammonia from their bodies by direct excretion. Ammonia even at dilute concentrations is highly toxic to aquatic animals, and for this reason it is classified as *dangerous for the environment*.

Ammonia is a constituent of tobacco smoke.

Coking Wastewater

Ammonia is present in coking wastewater streams, as a liquid by-product of the production of coke from coal. In some cases, the ammonia is discharged to the marine environment where it acts as a pollutant. The Whyalla steelworks in South Australia is one example of a coke-producing facility which discharges ammonia into marine waters.

Aquaculture

Ammonia toxicity is believed to be a cause of otherwise unexplained losses in fish hatcheries. Excess ammonia may accumulate and cause alteration of metabolism or increases in the body pH of the exposed organism. Tolerance varies among fish species. At lower concentrations, around 0.05 mg/L, un-ionised ammonia is harmful to fish species and can result in poor growth and feed conversion rates, reduced fecundity and fertility and increase stress and susceptibility to bacterial infections and diseases. Exposed to excess ammonia, fish may suffer loss of equilibrium, hyper-excitability, increased respiratory activity and oxygen uptake and increased heart rate. At concentrations exceeding 2.0 mg/L, ammonia causes gill and tissue damage, extreme lethargy, convulsions, coma, and death. Experiments have shown that the lethal concentration for a variety of fish species ranges from 0.2 to 2.0 mg/l.

During winter, when reduced feeds are administered to aquaculture stock, ammonia levels can be higher. Lower ambient temperatures reduce the rate of algal photosynthesis so less ammonia is removed by any algae present. Within an aquaculture environment, especially at large scale, there is no fast-acting remedy to elevated ammonia levels. Prevention rather than correction is recommended to reduce harm to farmed fish and in open water systems, the surrounding environment.

Storage Information

Similar to propane, anhydrous ammonia boils below room temperature when at atmospheric pressure. A storage vessel capable of 250 psi (1.7 MPa) is suitable to contain the liquid. Ammonium compounds should never be allowed to come in contact with bases (unless in an intended and contained reaction), as dangerous quantities of ammonia gas could be released.

Household use

Solutions of ammonia (5–10% by weight) are used as household cleaners, particularly for glass. These solutions are irritating to the eyes and mucous membranes (respiratory and digestive tracts), and to a lesser extent the skin. Caution should be used that the chemical is never mixed into any liquid containing bleach, as a poisonous gas may result. Mixing with chlorine-containing products or strong oxidants, such as household bleach, can lead to hazardous compounds such as chloramines.

Laboratory use of Ammonia Solutions

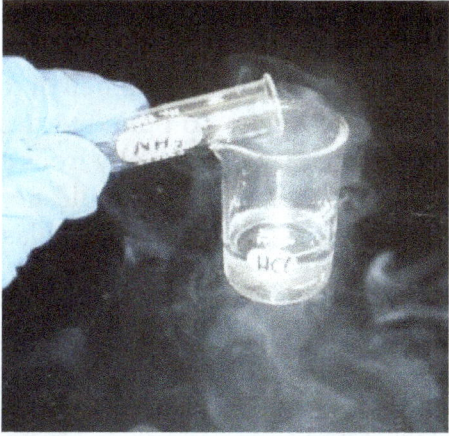

Hydrochloric acid sample releasing HCl fumes, which are reacting with ammonia fumes to produce a white smoke of ammonium chloride.

The hazards of ammonia solutions depend on the concentration: "dilute" ammonia solutions are usually 5–10% by weight (<5.62 mol/L); "concentrated" solutions are usually prepared at >25% by weight. A 25% (by weight) solution has a density of 0.907 g/cm^3, and a solution that has a lower density will be more concentrated. The European Union classification of ammonia solutions is given in the table.

Concentration by weight (w/w)	Molarity	Concentration mass/volume (w/v)	Classification	R-Phrases
5–10%	2.87–5.62 mol/L	48.9–95.7 g/L	Irritant (Xi)	R36/37/38
10–25%	5.62–13.29 mol/L	95.7–226.3 g/L	Corrosive (C)	R34
>25%	>13.29 mol/L	>226.3 g/L	Corrosive (C) Dangerous for the environment (N)	R34, R50

S-Phrases: (S1/2), S16, S36/37/39, S45, S61.

The ammonia vapour from concentrated ammonia solutions is severely irritating to the eyes and the respiratory tract, and these solutions should only be handled in a fume hood. Saturated solutions can develop a significant pressure inside a closed bottle in warm weather, and the bottle should be opened with care; this is not usually a problem for 25% ("0.900") solutions.

Ammonia solutions should not be mixed with halogens, as toxic and/or explosive products are formed. Prolonged contact of ammonia solutions with silver, mercury or iodide salts can also lead to explosive products: such mixtures are often formed in qualitative inorganic analysis, and should be lightly acidified but not concentrated (<6% w/v) before disposal once the test is completed.

Laboratory use of Anhydrous Ammonia (Gas or Liquid)

Anhydrous ammonia is classified as toxic (T) and dangerous for the environment (N). The gas is flammable (autoignition temperature: 651 °C) and can form explosive mixtures with air (16–25%). The permissible exposure limit (PEL) in the United States is 50 ppm (35 mg/m^3), while the IDLH concentration is estimated at 300 ppm. Repeated exposure to ammonia lowers the sensitivity to the smell of the gas: normally the odour is detectable at concentrations of less than 50 ppm, but desensitised individuals may not detect it even at concentrations of 100 ppm. Anhydrous ammonia corrodes copper- and zinc-containing alloys, and so brass fittings should not be used for handling the gas. Liquid ammonia can also attack rubber and certain plastics.

Ammonia reacts violently with the halogens. Nitrogen triiodide, a primary high explosive, is formed when ammonia comes in contact with iodine. Ammonia causes the explosive polymerisation of ethylene oxide. It also forms explosive fulminating compounds with compounds of gold, silver, mercury, germanium or tellurium, and with stibine. Violent reactions have also been reported with acetaldehyde, hypochlorite solutions, potassium ferricyanide and peroxides.

Liquid Ammonia as a Solvent

Liquid ammonia is the best-known and most widely studied nonaqueous ionising solvent. Its most conspicuous property is its ability to dissolve alkali metals to form highly coloured, electrically conductive solutions containing solvated electrons. Apart from these remarkable solutions, much of the chemistry in liquid ammonia can be classified by analogy with related reactions in aqueous solutions. Comparison of the physical properties of NH_3 with those of water shows NH_3 has the lower melting point, boiling point, density, viscosity, dielectric constant and electrical conductivity; this is due at least in part to the weaker hydrogen bonding in NH_3 and because such bonding cannot form cross-linked networks, since each NH_3 molecule has only one lone pair of electrons compared with two for each H_2O molecule. The ionic self-dissociation constant of liquid NH_3 at −50 °C is about 10^{-33} mol²·l⁻².

Solubility of Salts

	Solubility (g of salt per 100 g liquid NH_3)
Ammonium acetate	253.2
Ammonium nitrate	389.6
Lithium nitrate	243.7
Sodium nitrate	97.6
Potassium nitrate	10.4
Sodium fluoride	0.35
Sodium chloride	157.0
Sodium bromide	138.0
Sodium iodide	161.9
Sodium thiocyanate	205.5

Liquid ammonia is an ionising solvent, although less so than water, and dissolves a range of ionic compounds, including many nitrates, nitrites, cyanides, thiocyanates, metal cyclopentadienyl complexes and metal bis(trimethylsilyl)amides. Most ammonium salts are soluble and act as acids in liquid ammonia solutions. The solubility of halide salts increases from fluoride to iodide. A saturated solution of ammonium nitrate (Divers' solution, named after Edward Divers) contains 0.83 mol solute per mole of ammonia and has a vapour pressure of less than 1 bar even at 25 °C (77 °F).

Solutions of Metals

Liquid ammonia will dissolve the alkali metals and other electropositive metals such as magnesium, calcium, strontium, barium, europium and ytterbium. At low concentrations (<0.06 mol/l), deep blue solutions are formed: these contain metal cations and solvated electrons, free electrons that are surrounded by a cage of ammonia molecules.

These solutions are very useful as strong reducing agents. At higher concentrations, the solutions are metallic in appearance and in electrical conductivity. At low temperatures, the two types of solution can coexist as immiscible phases.

Redox properties of liquid ammonia

	$E°$ (V, ammonia)	$E°$ (V, water)
$Li^+ + e^- \rightleftharpoons Li$	−2.24	−3.04
$K^+ + e^- \rightleftharpoons K$	−1.98	−2.93
$Na^+ + e^- \rightleftharpoons Na$	−1.85	−2.71
$Zn^{2+} + 2e^- \rightleftharpoons Zn$	−0.53	−0.76
$NH_4^+ + e^- \rightleftharpoons \frac{1}{2} H_2 + NH_3$	0.00	—
$Cu^{2+} + 2e^- \rightleftharpoons Cu$	+0.43	+0.34
$Ag^+ + e^- \rightleftharpoons Ag$	+0.83	+0.80

The range of thermodynamic stability of liquid ammonia solutions is very narrow, as the potential for oxidation to dinitrogen, $E°$ ($N_2 + 6NH_4^+ + 6e^- \rightleftharpoons 8NH_3$), is only +0.04 V. In practice, both oxidation to dinitrogen and reduction to dihydrogen are slow. This is particularly true of reducing solutions: the solutions of the alkali metals mentioned above are stable for several days, slowly decomposing to the metal amide and dihydrogen. Most studies involving liquid ammonia solutions are done in reducing conditions; although oxidation of liquid ammonia is usually slow, there is still a risk of explosion, particularly if transition metal ions are present as possible catalysts.

Ammonia's role in biological systems and human disease

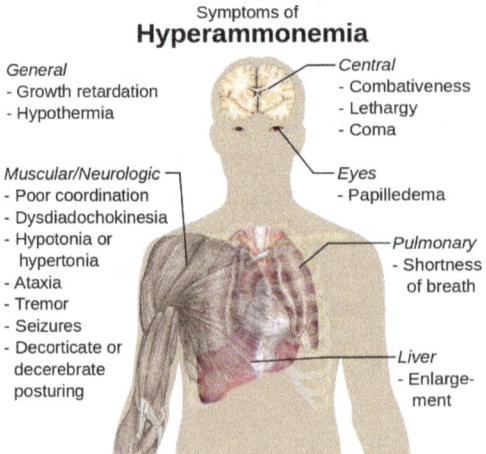

Main symptoms of hyperammonemia (ammonia reaching toxic concentrations).

Ammonia is both a metabolic waste and a metabolic input throughout the biosphere. It is an important source of nitrogen for living systems. Although atmospheric nitrogen abounds (more than 75%), few living creatures are capable of using this atmospheric nitrogen in its diatomic form, N_2 gas. Therefore, nitrogen fixation is required for the synthesis of amino acids, which are the building blocks of protein. Some plants rely on ammonia and other nitrogenous wastes incorporated into the soil by decaying matter. Others, such as nitrogen-fixing legumes, benefit from symbiotic relationships with rhizobia that create ammonia from atmospheric nitrogen.

Biosynthesis

In certain organisms, ammonia is produced from atmospheric nitrogen by enzymes called nitrogenases. The overall process is called nitrogen fixation. Intense effort has been directed toward understanding the mechanism of biological nitrogen fixation; the scientific interest in this problem is motivated by the unusual structure of the active site of the enzyme, which consists of an Fe_7MoS_9 ensemble.

Ammonia is also a metabolic product of amino acid deamination catalyzed by enzymes such as glutamate dehydrogenase 1. Ammonia excretion is common in aquatic animals. In humans, it is quickly converted to urea, which is much less toxic, particularly less basic. This urea is a major component of the dry weight of urine. Most reptiles, birds, insects, and snails excrete uric acid solely as nitrogenous waste.

In Physiology

Ammonia also plays a role in both normal and abnormal animal physiology. It is biosynthesised through normal amino acid metabolism and is toxic in high concentrations. The liver converts ammonia to urea through a series of reactions known as the urea cycle. Liver dysfunction, such as that seen in cirrhosis, may lead to elevated amounts of ammonia in the blood (hyperammonemia). Likewise, defects in the enzymes responsible for the urea cycle, such as ornithine transcarbamylase, lead to hyperammonemia. Hyperammonemia contributes to the confusion and coma of hepatic encephalopathy, as well as the neurologic disease common in people with urea cycle defects and organic acidurias.

Ammonia is important for normal animal acid/base balance. After formation of ammonium from glutamine, α-ketoglutarate may be degraded to produce two molecules of bicarbonate, which are then available as buffers for dietary acids. Ammonium is excreted in the urine, resulting in net acid loss. Ammonia may itself diffuse across the renal tubules, combine with a hydrogen ion, and thus allow for further acid excretion.

Excretion

Ammonium ions are a toxic waste product of metabolism in animals. In fish and aquatic invertebrates, it is excreted directly into the water. In mammals, sharks, and amphibians, it is converted in the urea cycle to urea, because it is less toxic and can be stored more efficiently. In birds, reptiles, and terrestrial snails, metabolic ammonium is converted into uric acid, which is solid, and can therefore be excreted with minimal water loss.

Reference ranges for blood tests, comparing blood content of ammonia (shown in yellow near middle) with other constituents

In Astronomy

Ammonia has been detected in the atmospheres of the gas giant planets, including Jupiter, along with other gases like methane, hydrogen, and helium. The interior of Saturn may include frozen crystals of ammonia. It is naturally found on Deimos and Phobos – the two moons of Mars.

Ammonia occurs in the atmospheres of the outer gas planets such as Jupiter (0.026% ammonia) and Saturn (0.012% ammonia).

Interstellar Space

Ammonia was first detected in interstellar space in 1968, based on microwave emissions from the direction of the galactic core. This was the first polyatomic molecule to be so detected. The sensitivity of the molecule to a broad range of excitations and the ease with which it can be observed in a number of regions has made ammonia one of the most important molecules for studies of molecular clouds. The relative intensity of the ammonia lines can be used to measure the temperature of the emitting medium.

The following isotopic species of ammonia have been detected:

NH_3, $^{15}NH_3$, NH_2D, NHD_2, and ND_3

The detection of triply deuterated ammonia was considered a surprise as deuterium is relatively scarce. It is thought that the low-temperature conditions allow this molecule to survive and accumulate.

Since its interstellar discovery, NH_3 has proved to be an invaluable spectroscopic tool in the study of the interstellar medium. With a large number of transitions sensitive to a wide range of excitation conditions, NH_3 has been widely astronomically detected – its detection has been reported in hundreds of journal articles.

Interstellar Formation Mechanisms

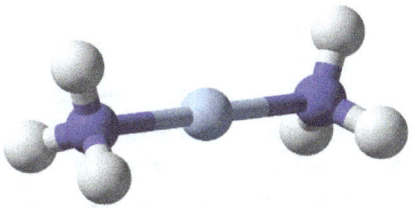

Ball-and-stick model of the diamminesilver(I) cation, $[Ag(NH_3)_2]^+$

The interstellar abundance for ammonia has been measured for a variety of environments. The $[NH_3]/[H_2]$ ratio has been estimated to range from 10^{-7} in small dark clouds up to 10^{-5} in the dense core of the Orion Molecular Cloud Complex. Although a total of 18 total production routes have been proposed, the principal formation mechanism for interstellar NH_3 is the reaction:

$NH_4^+ + e^- \rightarrow NH_3 + H\cdot$

The rate constant, k, of this reaction depends on the temperature of the environment, with a value of 5.2×10^{-6} at 10 K. The rate constant was calculated from the formula $k = a(T / 300)^B$. For the primary formation reaction, $a = 1.05\times10^{-6}$ and $B = -0.47$. Assuming an NH_4^+ abundance of 3×10^{-7} and an electron abundance of 10^{-7} typical of molecular clouds, the formation will proceed at a rate of 1.6×10^{-9} $cm^{-3}s^{-1}$ in a molecular cloud of total density 10^5 cm^{-3}.

All other proposed formation reactions have rate constants of between 2 and 13 orders of magnitude smaller, making their contribution to the abundance of ammonia relatively insignificant. As an example of the minor contribution other formation reactions play, the reaction:

$H_2 + NH_2 \rightarrow NH_3 + H$

has a rate constant of 2.2×10^{-15}. Assuming H_2 densities of 10^5 and $[NH_2]/[H_2]$ ratio of 10^{-7}, this reaction proceeds at a rate of 2.2×10^{-12}, more than 3 orders of magnitude slower than the primary reaction above.

Some of the other possible formation reactions are:

$$H^- + NH_4^+ \rightarrow NH_3 + H_2$$

$$PNH_3^+ + e^- \rightarrow P + NH_3$$

Interstellar Destruction Mechanisms

There are 113 total proposed reactions leading to the destruction of NH_3. Of these, 39 were tabulated in extensive tables of the chemistry among C, N, and O compounds. A review of interstellar ammonia cites the following reactions as the principal dissociation mechanisms:

$$NH3 + H3+ \rightarrow NH4+ + H2$$

$$NH3 + HCO+ \rightarrow NH4+ + CO$$

with rate constants of 4.39×10^{-9} and 2.2×10^{-9}, respectively. The above equations (1, 2) run at a rate of 8.8×10^{-9} and 4.4×10^{-13}, respectively. These calculations assumed the given rate constants and abundances of $[NH_3]/[H_2] = 10^{-5}$, $[H_3^+]/[H_2] = 2\times10^{-5}$, $[HCO^+]/[H_2] = 2\times10^{-9}$, and total densities of $n = 10^5$, typical of cold, dense, molecular clouds. Clearly, between these two primary reactions, equation $(NH3 + H3+ \rightarrow NH4+ + H2)$ is the dominant destruction reaction, with a rate $\sim$10,000 times faster than equation $(NH3 + HCO+ \rightarrow NH4+ + CO)$. This is due to the relatively high abundance of H_3^+.

Single Antenna Detections

Radio observations of NH_3 from the Effelsberg 100-m Radio Telescope reveal that the ammonia line is separated into two components – a background ridge and an unresolved core. The background corresponds well with the locations previously detected CO. The 25 m Chilbolton telescope in England detected radio signatures of ammonia in H II regions, HNH_2O masers, H-H objects, and other objects associated with star formation. A comparison of emission line widths indicates that turbulent or systematic velocities do not increase in the central cores of molecular clouds.

Microwave radiation from ammonia was observed in several galactic objects including W3(OH), Orion A, W43, W51, and five sources in the galactic centre. The high detection rate indicates that this is a common molecule in the interstellar medium and that high-density regions are common in the galaxy.

Interferometric Studies

VLA observations of NH_3 in seven regions with high-velocity gaseous outflows revealed condensations of less than 0.1 pc in L1551, S140, and Cepheus A. Three individual condensations were detected in Cepheus A, one of them with a highly elongated shape. They may play an important role in creating the bipolar outflow in the region.

Extragalactic ammonia was imaged using the VLA in IC 342. The hot gas has temperatures above 70 K, which was inferred from ammonia line ratios and appears to be closely associated with the innermost portions of the nuclear bar seen in CO. NH_3 was also monitored by VLA toward a sample of four galactic ultracompact HII regions: G9.62+0.19, G10.47+0.03, G29.96-0.02, and G31.41+0.31. Based upon temperature and density diagnostics, it is concluded that in general such clumps are probably the sites of massive star formation in an early evolutionary phase prior to the development of an ultracompact HII region.

Infrared Detections

Absorption at 2.97 micrometres due to solid ammonia was recorded from interstellar grains in the Becklin-Neugebauer Object and probably in NGC 2264-IR as well. This detection helped explain the physical shape of previously poorly understood and related ice absorption lines.

A spectrum of the disk of Jupiter was obtained from the Kuiper Airborne Observatory, covering the 100 to 300 cm^{-1} spectral range. Analysis of the spectrum provides information on global mean properties of ammonia gas and an ammonia ice haze.

A total of 149 dark cloud positions were surveyed for evidence of 'dense cores' by using the (J,K) = (1,1) rotating inversion line of NH_3. In general, the cores are not spherically shaped, with aspect ratios ranging from 1.1 to 4.4. It is also found that cores with stars have broader lines than cores without stars.

Ammonia has been detected in the Draco Nebula and in one or possibly two molecular clouds, which are associated with the high-latitude galactic infrared cirrus. The finding is significant because they may represent the birthplaces for the Population I metallicity B-type stars in the galactic halo that could have been borne in the galactic disk.

Observations of Nearby Dark Clouds

By balancing and stimulated emission with spontaneous emission, it is possible to construct a relation between excitation temperature and density. Moreover, since the transitional levels of ammonia can be approximated by a 2-level system at low temperatures, this calculation is fairly simple. This premise can be applied to dark clouds, regions suspected of having extremely low temperatures and possible sites for future star formation. Detections of ammonia in dark clouds show very narrow lines—indicative not only of low temperatures, but also of a low level of inner-cloud turbulence. Line ratio calculations provide a measurement of cloud temperature that is independent of previous CO observations. The ammonia observations were consistent with CO measurements of rotation temperatures of ~10 K. With this, densities can be determined, and have been calculated to range between 10^4 and 10^5 cm^{-3} in dark clouds. Mapping of NH$_3$ gives typical clouds sizes of 0.1 pc and masses near 1 solar mass. These cold, dense cores are the sites of future star formation.

UC HII Regions

Ultra-compact HII regions are among the best tracers of high-mass star formation. The dense material surrounding UCHII regions is likely primarily molecular. Since a complete study of massive star formation necessarily involves the cloud from which the star formed, ammonia is an invaluable tool in understanding this surrounding molecular material. Since this molecular material can be spatially resolved, it is possible to constrain the heating/ionising sources, temperatures, masses, and sizes of the regions. Doppler-shifted velocity components allow for the separation of distinct regions of molecular gas that can trace outflows and hot cores originating from forming stars.

Extragalactic Detection

Ammonia has been detected in external galaxies, and by simultaneously measuring several lines, it is possible to directly measure the gas temperature in these galaxies. Line ratios imply that gas temperatures are warm (~50 K), originating from dense clouds with sizes of tens of pc. This picture is consistent with the picture within our Milky Way galaxy—hot dense molecular cores form around newly forming stars embedded in larger clouds of molecular material on the scale of several hundred pc (giant molecular clouds; GMCs).

Aluminium Hydroxide

Aluminium hydroxide is a inorganic basic compound used as intermediary in organic synthesis and as additive in pharmaceutical and fine chemical industries.

Formula and structure: The aluminium hydroxide chemical formula is Al(OH)3 and its molar mass is 78.00 g mol-1. The molecule is formed by the aluminium cation Al+3 and three hydroxyl anions CO3-2. The structure of the aluminium hydroxide lattice depends on the mineral from it is extracted because the ions show different arrangements. Most of the lattice are hexagonal or orthorhombic. Its chemical structure can be written as below, in the common representations used for organic molecules.

$$OH^-$$
$$Al^{3+}$$
$$OH^- \quad OH^-$$

Aluminium hydroxide

Occurrence: Aluminium hydroxide, similar to other metals carbonates, hydroxides and sulfates, is found in mineral ores of gibbsites, bayertute, doyleite and strandite.

Preparation: Although aluminium hydroxide is largely found in many geological systems in nature, it is mostly produced by the Bayer and sintering processes to obtain alumina from the mineral bauxite. Thus, 97% of the world aluminium hydroxide is obtaining through the treatment of bauxite with caustic soda yielding sodium aluminate, which is decomposes by stirring to obtain an aluminium hydroxide precipitate:

$$NaAl(OH)_4 \longrightarrow Al(OH)_3 + NaOH$$

Other processes to obtain aluminium hydroxide are the hydrothermal technique, the micro-emulsion or the Sol-gel. These methods have the advantage of producing an compound with a higher level of purity.

Physical properties: Aluminium hydroxide is an odorless, white amorphouse solid. Its density is 2.42 g mL^{-1}. Aluminium hydroxide melting point is 300 $^\circ$C. It is insoluble in water and ethanol, but soluble in acids and alkalis solutions.

Chemical properties: Aluminium hydroxide is an amphoteric compound, which means that the substance presents basic or acid characteristics. Consequently, the aluminium hydroxide is soluble in both: acids (reaction I) or alkalis (reaction II) solutions:

$$Al(OH)_3 + 3H^+ \longrightarrow Al^{+3} + H_2O\,(I)$$

$$Al(OH)_3 + OH^- \longrightarrow AlO_2^- + H_2O\,(II)$$

Uses

One of the major uses of aluminium hydroxide is as a feedstock for the manufacture of other aluminium compounds: speciality calcined aluminas, aluminium sulfate, polyaluminium chloride, aluminium chloride, zeolites, sodium aluminate, activated alumina, and aluminium nitrate.

Freshly precipitated aluminium hydroxide forms gels, which are the basis for the application of aluminium salts as flocculants in water purification. This gel crystallizes with time. Aluminium hydroxide gels can be dehydrated (e.g. using water-miscible non-aqueous solvents like ethanol) to form an amorphous aluminium hydroxide powder, which is readily soluble in acids. Aluminium hydroxide powder which has been heated to an elevated temperature under carefully controlled conditions is known as activated alumina and is used as a desiccant, as an adsorbent in gas purification, as a Claus catalyst support for water purification, and as an adsorbent for the catalyst during the manufacture of polyethylene by the Sclairtech process.

Fire Retardant

Aluminium hydroxide also finds use as a fire retardant filler for polymer applications in a similar way to magnesium hydroxide and mixtures of huntite and hydromagnesite. It decomposes at about 180 °C (356 °F), absorbing a considerable amount of heat in the process and giving off water vapour. In addition to behaving as a fire retardant, it is very effective as a smoke suppressant in a wide range of polymers, most especially in polyesters, acrylics, ethylene vinyl acetate, epoxies, PVC and rubber.

Pharmaceutical

Under the generic name "algeldrate", aluminium hydroxide is used as an antacid in humans and animals (mainly cats and dogs). It is preferred over other alternatives such as sodium bicarbonate because $Al(OH)_3$, being insoluble, does not increase the pH of stomach above 7 and hence, does not trigger secretion of excess acid by the stomach. Brand names include Alu-Cap, Aludrox, Gaviscon or Pepsamar. It reacts with excess acid in the stomach, reducing the acidity of the stomach content, which may relieve the symptoms of ulcers, heartburn or dyspepsia. Such products can cause constipation, because the aluminium ions inhibit the contractions of smooth muscle cells in the gastrointestinal tract, slowing peristalsis and lengthening the time needed for stool to pass through the colon. Some such products (such as Maalox) are formulated to minimize such effects through the inclusion of equal concentrations of magnesium hydroxide or magnesium carbonate, which have counterbalancing laxative effects.

This compound is also used to control hyperphosphatemia (elevated phosphate, or phosphorus, levels in the blood) in people and animals suffering from kidney failure. Normally, the kidneys filter excess phosphate out from the blood, but kidney failure can cause phosphate to accumulate. The aluminium salt, when ingested, binds to phosphate in the intestines and reduce the amount of phosphorus that can be absorbed.

Precipitated aluminium hydroxide is included as an adjuvant in some vaccines (e.g. anthrax vaccine). One of the well-known brands of aluminium hydroxide adjuvant is Alhydrogel, made by Brenntag Biosector. Since it absorbs protein well, it also functions to stabilize vaccines by preventing the proteins in the vaccine from precipitating or sticking to the walls of the container during storage. Aluminium hydroxide is sometimes mistakenly called "alum", which properly refers to aluminium potassium sulfate.

Vaccine formulations containing aluminium hydroxide stimulate the immune system by inducing the release of uric acid, an immunological danger signal. This strongly attracts certain types of monocytes which differentiate into dendritic cells. The dendritic cells pick up the antigen, carry it to lymph nodes, and stimulate T cells and B cells. It appears to contribute to induction of a good Th2 response, so is useful for immunizing against pathogens that are blocked by antibodies. However, it has little capacity to stimulate cellular (Th1) immune responses, important for protection against many pathogens, nor is it useful when the antigen is peptide-based.

Potential Adverse Effects

In the 1960s and 1970s it was speculated that aluminium was related to various neurological disorders, including Alzheimer's disease. Since then, multiple epidemiological studies have found no connection between exposure to aluminium and neurological disorders.

Barium Chlorate

Barium Chlorate is a white crystalline solid. Forms very flammable mixtures with combustible materials. Mixtures may be ignited by friction and may be explosive if the combustible material is finely divided. Contact with concentrated sulfuric acid solutions may cause fires or explosions. May spontaneously decompose and ignite when mixed with ammonium salt. May explode under prolonged exposure to heat or fire. Used in explosives and pyrotechnics, in dyeing textiles, and to make other chlorates.

The ability of a molecule to act as an oxidizer is at best only vaguely related to the count of oxygens per molecule.

Barium chlorate has a formula weight of 304.23. Sure, there are 6 oxygens, but there is a lot of weight in the Barium atoms. A 100 gram sample of Barium Chlorate contains 31.55 grams of oxygen atoms. Potassium chlorate has a formula weight of 122.55, with 3 oxygens. This means that a 100 gram sample contains 39.17% oxygen. In fact, on a weight/weight basis, there is more oxygen in a gram of Potassium Chlorate than in a gram of Barium Chlorate.

Next - Barium surprisingly has essentially the same radius as Potassium. The radii of Barium 2+ and Potassium + are 1.34 Ang, and 1.33 Ang respectively.

So, the difference in the crystal lattice stability cannot be ascribed to the difference in size. Then we get to the greater conundrum, heats of formation. Barium Chlorate is actually.

$Ba(ClO_3)_2$ by Electrolyses

Barium chlorate is used for making vivid green colour in pyrotechnics. If you make it by double decomposition with Barium chloride and Sodium chlorate it will be difficult to remove all of the sodium ion from the product and the yellow colour that sodium gives will effect the vivid green colour that you desired. Some have suggested using Calcium chlorate instead of Sodium chlorate.

It can be produced easily enough by electrolyses but the current efficiency is lower than with other chlorates.

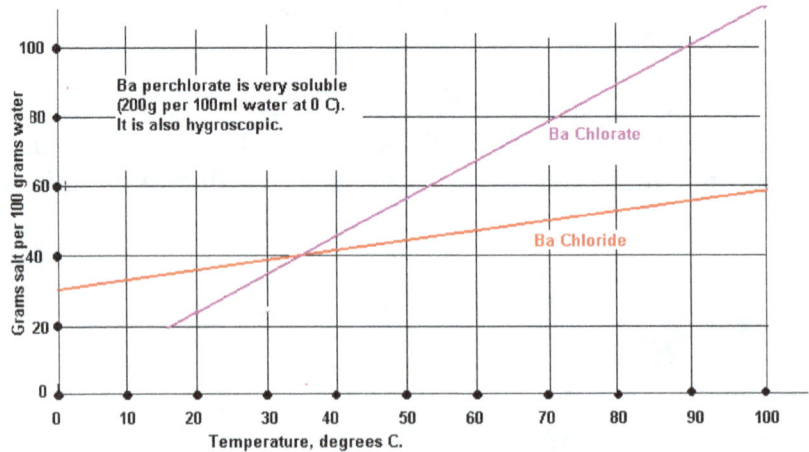

Some properties of Ba chloride and chlorate		
	$BaCl_2 : 2H_2O$	$Ba(ClO_3)_2$
Molecular weight	244	304

The solubilities of the chlorate and the chloride are not very far apart and it is not possible to get a large crop of pure crystals out of the electrolytic cell. This is not a problem as you can take a small crop and recycle the rest. About 40g chloride per 100ml water is put into the cell and the cell is run for the required run time. It should be noted that the chloride has water of crystallization and has the formula BaCl2:2H2O. The current efficiency will be about 30 - 40% without pH control. The cell liquor is evaporated until crystals form and a crop of Ba chlorate removed.

Properties

Chemical

Barium chlorate is is an extremely powerful oxidizer.

If sulfuric acid is added to barium chlorate, chloric acid is formed:

$$Ba(ClO_3)_2 + H_2SO_4 \longrightarrow 2\,HClO_3 + BaSO_4\,(\text{precipitates})$$

Since barium sulfate is very insoluble, this reaction is useful to obtain chloric acid of high purity: just filter the solution.

If the experiment is run dry (i.e. in a non-aqueous environment), the pure, very unstable chloric acid decomposes immediately to perchloric acid and chlorine dioxide. The latter will spontaneously ignite any combustible material (sugar, paper, dust).

However, doing this reaction aqueously lets one synthesize stable solutions of chloric acid.

Barium chlorate always gives the green color to the flame if it is a component of a fuel-oxidizer mix. This is its main use in pyrotechnics.

Physical

Barium chlorate is a transparent to white salt that are poorly soluble in water and glycerol. It is even less soluble in cold water than potassium chlorate, which allows for easy conversion between two salts: in a cooled solution of $BaCl2$ and potassium chlorate, the double displacement reaction proceeds almost fully.

Availability

It can be found in some pyrotechnics.

Preparation

The common precursor chem to barium chlorate is potassium chlorate, in form of the warm solution. You will also need ammonia and tartric acid. Bubble ammonia through a solution of tartaric acid until you get ammonium tartrate, or mix aqueous ammonia (fresh, so you know the concentration) with tartric acid. Add both solutions to barium carbonate and boil until ammonia and carbon dioxide stop emerging. You will notice that the insoluble, chalky barium carbonate dissolves. Ammonia will emerge and briefly liberate chloric acid, which will react with barium carbonate; the insoluble potassium bitartrate will precipitate.

$$BaCO_3 + 2\,(NH_4)_2\,C_6H_2O_6 + 2\,KClO_3 \longrightarrow Ba(ClO_3)_2 + H_2O + CO_{2(gas)}$$
$$+ 2\,KHC_6H_2O_{6(prec.)} + 4\,NH_{3(gas)}$$

The same synthesis can be done with any other acid with a soluble ammonium salt and an insoluble potassium salt, such as hexanitritocobaltic. But tartric is just cheaper.

Do not use barium hydroxide, it will lead to an undesirable side reaction: formation of barium tartrate!

Reactions

Synthesis

Barium chlorate can be produced through a double replacement reaction between solutions of barium chloride and sodium chlorate:

$$BaCl_2 + 2\,NaClO_3 \rightarrow Ba(ClO_3)_2 + 2\,NaCl$$

On concentrating and chilling the resulting mixture, barium chlorate precipitates. This is perhaps the most common preparation, exploiting the lower solubility of barium chlorate compared to sodium chlorate.

The above method does result in some sodium contamination, which is undesirable for pyrotechnic purposes, where the strong yellow of sodium can easily overpower the green of barium. Sodium-free barium chlorate can be produced directly through electrolysis:

$$BaCl_2 + 6\,H_2O \rightarrow Ba(ClO_3)_2 + 6\,H_2$$

It can also be produced by the reaction of barium carbonate with boiling ammonium chlorate solution:

$$2\,NH_4ClO_3 + BaCO_3 + Q \rightarrow Ba(ClO_3)_2 + 2\,NH_3 + H_2O + CO_2$$

The reaction initially produces barium chlorate and ammonium carbonate; boiling the solution decomposes the ammonium carbonate and drives off the resulting ammonia and carbon dioxide, leaving only barium chlorate in solution.

The green seen in this firework is produced by barium chlorate and barium nitrate

Decomposition

When exposed to heat, barium chlorate alone will decompose to barium chloride and oxygen:

$$Ba(ClO_3)_2 \rightarrow BaCl_2 + 3O_2$$

Chloric acid

Barium chlorate is used to produce chloric acid, the formal precursor to all chlorate salts, through its reaction with dilute sulfuric acid, which results in a solution of chloric acid and insoluble barium sulfate precipitate:

$$Ba(ClO_3)_2 + H_2SO_4 \rightarrow 2HClO_3 + BaSO_4$$

Both the chlorate and the acid should be prepared as dilute solutions before mixing, such that the chloric acid produced is dilute, as concentrated solutions of chloric acid (above 30%) are unstable and prone to decompose, sometimes explosively.

Commercial uses

Fireworks

Barium chlorate, when burned with a fuel, produces a vibrant green light. Because it is an oxidizer, a chlorine donor, and contains a metal, this compound produces a green color that is unparalleled. However, due to the instability of all chlorates to sulfur, acids, and ammonium ions, chlorates have been banned from use in class C fireworks in the United States. Therefore, more and more firework producers have begun to use more stable compound such as barium nitrate and barium carbonate.

Environmental Hazard

Barium chlorate is toxic to humans and can also harm the environment. It is very harmful to aquatic organisms if it is leached into bodies of water. Chemical spills of this compound, although not common, can harm entire ecosystems and should be prevented. It is necessary to dispose of this compound as hazardous waste. The Environmental Protection Agency (EPA) lists barium chlorate as hazardous.

Handling

Safety

When mixed with combustible materials, even those normally slightly flammable (such as dust and lint), it will burn vigorously in combination and the fires are extremely hard to put out, as the chlorate provides the oxygen for the fire. Sulfur and red phosphorus, should be avoided in pyrotechnic compositions containing barium chlorate, as well as any acidic salts, as these mixtures are shock and friction sensitive and prone to spontaneous deflagration (in the safety head matches, such mixture is stabilized with glue).

Like all soluble barium compounds, this one is acutely toxic if ingested!

Storage

Barium chlorate should be stored in closed containers and away from any organic sources, as well as strong acidic vapors. Since it is not hygroscopic, it is not necessary to keep it air tight.

Disposal

Barium chlorate can be neutralized in two steps. First, with a reducing agent, such as sodium metabisulfite, sodium bisulfite, sodium sulfite. Then, the toxic barium remains, and this is neutralized with sulfuric acid. Using sulfites allows combining both steps, since they oxidize to sulfates and neutralize barium in their oxidized form.

Zirconium Tungstate

Zirconium Tungstate is a complex oxide that exhibits the unusual property of contracting, rather than expanding, as its temperature rises from near absolute zero to its decomposition-temperature near 1050°K (780°C). Throughout this range, Zirconium Tungstate has cubic symmetry, so that its thermal expansion is the same in any direction. Near 428°K (155°C), Zirconium Tungstate undergoes a second-order phase transformation to a disordered phase of higher symmetry, called β-Zirconium Tungstate to distinguish it from the α-phase, the form stable below 428°K. When exposed to pressure at room temperature, α-Zirconium Tungstate converts to a denser polymorph, called the γ-phase. This phase persists at atmospheric pressure, but reverts to the α-form when heated.

ZrW_2O_8 Crystal Structure and Origin of Negative Thermal Expansion (NTE)

Thermal expansion is described as the change of geometrical parameters, length and volume, of the materials with temperature. It can be positive as well as negative depending on the structure and strength of the type of bonds in a material. As indicated in table, ceramics have lower expansion coefficients than those of metals and polymers. Ceramics with low CTE also have relatively stronger interatomic forces. Since ceramics are brittle and may undergo fracture, they should have isotropic low thermal expansion to resist the thermal shocks.

CTE of a solid material is closely related to interatomic potentials. If the bonds are strong, interatomic potentials have smaller values, and narrow and symmetrical potential energy curve are obtained.

Therefore, vibration shows more harmonic behavior and so affects the interatomic distance less. As the harmonic vibrations are obtained, average distance between two atoms does not change much with the increase of temperature. However, interatomic potential functions are not harmonical in real cases. Therefore, average distances between two atoms increase with temperature, which results in PTE.

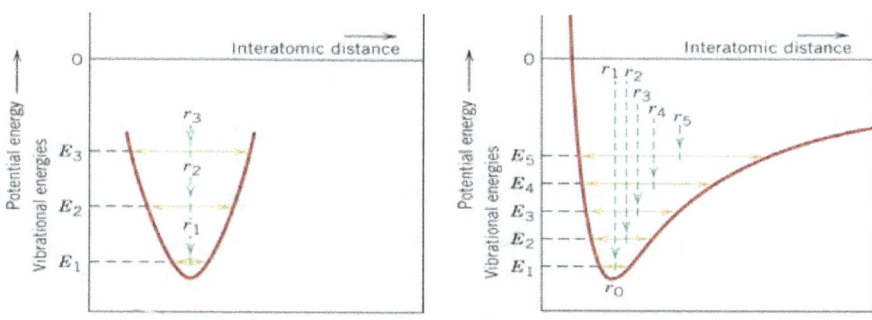

Plot of potential energy versus interatomic distance
(a) harmonic oscillator and (b) non-harmonic oscillator.

If the materials have stronger atomic interactions, therefore stronger bonds, in their crystal structures, they have generally low or NTE coefficients. Although stronger bonds affect the expansion coefficient of the materials, there are some additional mechanisms that play an important role on NTE mechanism for certain families of materials. Atomic vibrations of the materials occur mainly in two directions, longitudinal and transversal. These two mechanisms are presented in figure. If the materials show longitudinal vibration with temperature increase, the vertical distance between the two atoms increase, therefore PTE is observed in the material.

On the other hand, transversal vibration of the atoms may causes a decrease in distance between two atoms, so NTE can be observed.

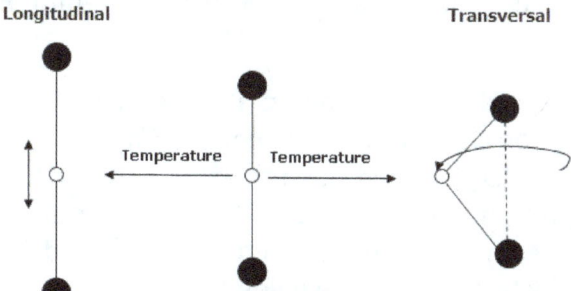

Longitudinal and transversal vibration modes

ZrW_2O_8, a NTE material over a wide range of temperature, belongs to $(A = Zr, Hf; X = W, Mo)$ W, Mo) family of materials. This family shows negative thermal expansion because of transversal vibration of oxygen atom in M-O-M (M: metal and O: oxygen) linkages, W-O-Zr in this case. Vibration of the oxygen atom is perpendicular to the strongly bonded W-O-Zr linkage, therefore the distance between W and Zr decreases. In figure below, global lattice shrinkage is illustrated. As the temperature increases (from Figure a to b) vibration of the oxygen atom and the angle θ will increase. Therefore crystal lattice will deform such that this results in an effective bond shortening in the unit cell. If this shortening is greater than the thermal expansion of individual polyhedral units, a net negative thermal expansion will be observed in the overall structure.

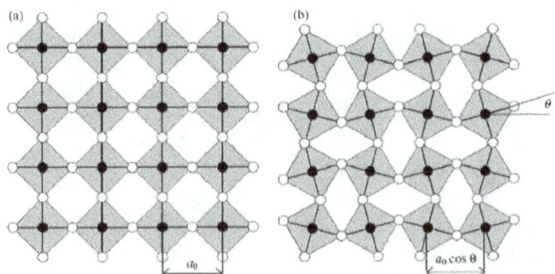

Schematic representation of global lattice shrinkage with temperature increase

There are some conditions to have this type of negative thermal expansion mechanism;

1. There should be a high M-O covalency $\left(M = W^{6+}, V^{5+}, Si^{4+}\right)$ because of the necessity of strong M-O bonds.

2. Coordinate oxygens should vibrate transversally.

3. There should be an open framework structure to support these transverse vibrational modes with low-energy.

4. All metal ions should be coordinated with oxygen atoms. Therefore, there should be no interstitial cations.

5. For lower symmetry and lower volume structures, displacive phase transitions are needed.

AX_2O_8 family of materials, and so ZrW_2O_8, meet these requirements and they show isotropic NTE.

ZrW_2O_8 was first synthesized by J. Graham et al. in 1959 by heating a mixture of ZrO_2 and WO_3 together at 12000 C followed by quenching. Cage structure of ZrW_2O_8 consists of ZrO_6 octahedra and WO_4 tetrahedra. In the structure ZrO_6 octahedrons share all corners with six WO4 tetrahedrons, whereas WO_4s share only 3 of its 4 oxygens with adjacent ZrO_6. Therefore, one oxygen atom remains unshared. The crystal structure at room temperature is illustrated in Figure below.

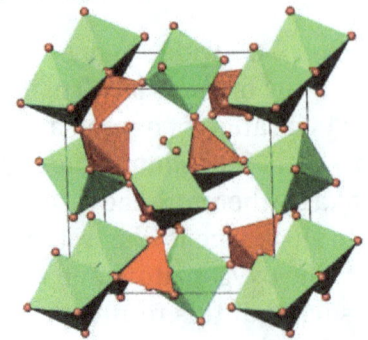

Crystal structure of ZrW_2O_8. WO_4 tetrahedrons and ZrO_6 octahedrons are shown in red and green, respectively. Oxygen atoms at the corners are shown as red spheres.

In the WO_4 tetrahedron, one oxygen atom is terminal and the arrangement of the WO_4 groups is such that pairs of tetrahedra are positioned along the main three-fold axis of the cubic unit cell with an asymmetric W···O−W bridge. Because of this geometry, the distance between W and terminal O is 1.7 Å, whereas the distance between this terminal O and adjacent W is 2.4 Å. Therefore, vibrations of the bridging oxygen atoms in the rigid open framework structure of the ZrW_2O_8 result in NTE. Additionally, atransversal vibration of O atoms in W-O-Zr linkages promotes NTE behavior. It was found that unit cell parameter is reduced from 9.3 Å to 8.8 Å without any deformation of the polyhedra.

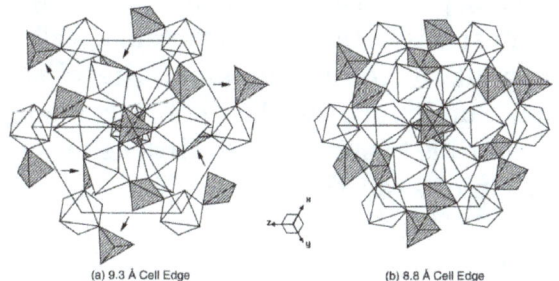

(a) 9.3 Å Cell Edge   (b) 8.8 Å Cell Edge

Idealized structures down the axis. The arrows in (a) show motion of the WO_4 tetrahedra (shaded) after, (b) tilting of ZrO_6 octahedra (unshaded).

Aluminum Oxide

Aluminum oxide, with the chemical formula Al_2O_3, is an amphoteric oxide and is commonly referred to as alumina. Corundum (α-aluminum oxide), emery, sapphire, amethyst, topaz, as well as many other names are reflecting its widespread occurrence in nature and industry. Corundum is the most common naturally occurring crystalline form of aluminum oxide. Rubies and sapphires are gem-quality forms of corundum, which owe their characteristic colors to trace impurities. Rubies are given their characteristic deep red color and their laser qualities by traces of chromium. Sapphires come in different colors given by various other impurities, such as iron and titanium.

Its most significant use is in the production of aluminum metal, although it is also used as an abrasive due to its hardness and as a refractory material due to its high melting point.

Physical Characteristics of Aluminum Oxide

General Properties

Aluminum oxide is a white powdery substance that has no odor. It is non-toxic, but airborne aluminum oxide dust can create industrial hazards, so wearing masks is recommended for prolonged exposure. Aluminum oxide is very heavy; a cube of aluminum oxide, 1 meter on a side, weighs about 7,200 lbs.

Industrial Properties

The aluminum oxide compound can be machined or molded into hard, wear-resistant materials suitable for usage in a variety of industrial roles. These include wire guides, machinery seals, metering devices and high temperature electrical insulators.

Chemical Properties

Aluminum oxide does not dissolve in water and has a very high melting point of 2,000 C or about 3,600 F. Its boiling point is an extremely high 5,400 F. The chemical formula combines two aluminum atoms to three oxygen atoms, which is expressed as Al_2O_3. It is an electrical resistor, unlike its cousin aluminum. The resistance level changes with the purity of the material. Aluminum oxide does not react readily with most materials, but it is higly reactive to chlorine trifluoride and ethylene oxide. Mixing aluminum oxide with either of these chemicals causes a fire.

Mechanical Properties

Aluminum oxide is a very hard material, almost to the level of diamonds, so it has excellent wear resistance properties. It has high corrosion endurance and high temperature stability, low thermal expansion and a favorable stiffness-to-weight ratio. Since aluminum oxide has an excellent electrical resistor, it is often used in capacitors as the dielectric, the part keeping charges in the device separated.

Production

The production of aluminum oxide is mainly from bauxite (the main aluminum ore), which is a mixture of various minerals including gibbsite $(Al(OH)_3)$, , boehmite $(\gamma - AlO(OH))$, and diaspore $\alpha - AlO(OH)$ along with impurities of iron oxides, quartz, and silicates.

Bauxite is purified by the Bayer process which is the principal industrial refining process. As bauxite contains only about 40 to 50% of alumina, the rest has been removed. This is achieved by washing bauxite with hot sodium hydroxide, which dissolves the alumina by converting it to aluminum hydroxide which forms a solution in a strong base:

$$Al_2O_3 + 2OH^- + 3H_2O \rightarrow 2[Al(OH)_4]^-$$

The other components of the bauxite do not dissolve and are filtered off (the residues usually form a red sludge which presents a disposal problem, since it contains, for example, arsen and cadmium). Next the solution is cooled which causes precipitation of a fluffy solid (aluminum hydroxide). The aluminum hydroxide is then heated to 1050°C which causes it to decompose into aluminum oxide and water:

$$2Al(OH)_3 \rightarrow Al_2O_3 + 3H_2O$$

The most common form of crystalline aluminium oxide is known as corundum, which is the thermodynamically stable form. The oxygen ions form a nearly hexagonal close-packed structure with the aluminium ions filling two-thirds of the octahedral interstices. Each Al^{3+} center is octahedral. In terms of its crystallography, corundum adopts a trigonal Bravais lattice with a space group of R-3c (number 167 in the International Tables). The primitive cell contains two formula units of aluminium oxide.

Aluminium oxide also exists in other phases, including the cubic γ and η phases, the monoclinic θ phase, the hexagonal χ phase, the orthorhombic κ phase and the δ phase that can be tetragonal or orthorhombic. Each has a unique crystal structure and properties. Cubic $γ-Al_2O_3$ has important technical applications. The so-called $β-Al_2O_3$ proved to be $NaAl_{11}O_{17}$.

Molten aluminium oxide near the melting temperature is roughly 2/3 tetrahedral (i.e. 2/3 of the Al are surrounded by 4 oxygen neighbors), and 1/3 5-coordinated, very little (<5%) octahedral Al-O is present. Around 80% of the oxygen atoms are shared among three or more Al-O polyhedra, and the majority of inter-polyhedral connections are corner-sharing, with the remaining 10–20% being edge-sharing. The breakdown of octahedra upon melting is accompanied by a relatively large volume increase (~20%), the density of the liquid close to its melting point is 2.93 g/cm³.

Applications

Over 90% of the aluminium oxide, normally termed Smelter Grade Alumina (SGA), produced is consumed for the production of aluminium, usually by the Hall–Héroult process. The remainder, normally called speciality alumina is used in a wide variety of applications which reflect its inertness, temperature resistance and electrical resistance.

- Filler:

  Being fairly chemically inert and white, aluminium oxide is a favored filler for plastics. Aluminium oxide is a common ingredient in sunscreen and is sometimes also present in cosmetics such as blush, lipstick, and nail polish.

- Glass:

  Many formulations of glass have aluminium oxide as an ingredient.

- Catalysis:

  Aluminium oxide catalyses a variety of reactions that are useful industrially. In its largest scale application, aluminium oxide is the catalyst in the Claus process for converting hydrogen sulfide waste gases into elemental sulfur in refineries. It is also useful for dehydration of alcohols to alkenes.

  Aluminium oxide serves as a catalyst support for many industrial catalysts, such as those used in hydrodesulfurization and some Ziegler-Natta polymerizations.

- Purification:

  Aluminium oxide is widely used to remove water from gas streams.

- Abrasive:

  Aluminium oxide is used for its hardness and strength. It is widely used as an abrasive, including as a much less expensive substitute for industrial diamond. Many types of sandpaper use aluminium oxide crystals. In addition, its low heat retention and low specific heat make it widely used in grinding operations, particularly cutoff tools. As the powdery abrasive mineral aloxite, it is a major component, along with silica, of the cue tip "chalk" used in billiards. Aluminium oxide powder is used in some CD/DVD polishing and scratch-repair kits. Its polishing qualities are also behind its use in toothpaste.

- Paint:

  Aluminium oxide flakes are used in paint for reflective decorative effects, such as in the automotive or cosmetic industries.

- Composite fiber:

  Aluminium oxide has been used in a few experimental and commercial fiber materials for high-performance applications (e.g., Fiber FP, Nextel 610, Nextel 720). Alumina nanofibers in particular have become a research field of interest.

- Personal armor:

  Some body armors utilize alumina ceramic plates, usually in combination with aramid or UHMWPE backing to achieve effectiveness against even most rifle threats. Alumina ceramic armor is readily available to most civilians in jurisdictions where it is legal, but is not considered military grade.

- Abrasion protection:

  Aluminium oxide can be grown as a coating on aluminium by anodizing or by plasma electrolytic oxidation. Both the hardness and abrasion-resistant characteristics of the coating originate from the high strength of aluminium oxide, yet the porous coating layer produced with conventional direct current anodizing procedures is within a 60-70 Rockwell hardness C range which is comparable only to hardened carbon steel alloys, but considerably inferior to the hardness of natural and synthetic corundum. Instead, with plasma electrolytic oxidation, the coating is porous only on the surface oxide layer while the lower oxide layers are much more compact than with standard DC anodizing procedures and present a higher crystallinity due to the oxide layers being remelted and densified to obtain α-Al2O3 clusters with much higher coating hardness values circa 2000 Vickers hardness.

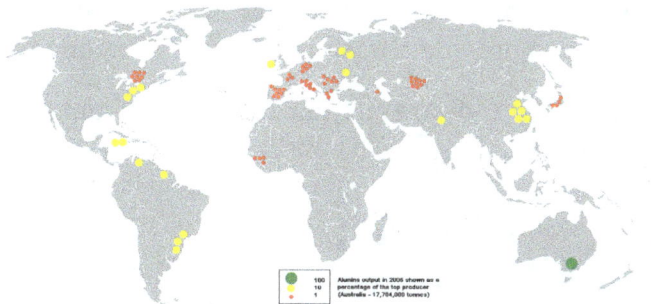

Aluminium oxide output in 2005

Alumina is used to manufacture tiles which are attached inside pulverized fuel lines and flue gas ducting on coal fired power stations to protect high wear areas. They are not suitable for areas with high impact forces as these tiles are brittle and susceptible to breakage.

- Other:

  In lighting, transparent aluminium oxide is used in some sodium vapor lamps. Aluminium oxide is also used in preparation of coating suspensions in compact fluorescent lamps.

In chemistry laboratories, aluminium oxide is a medium for chromatography, available in basic (pH 9.5), acidic (pH 4.5 when in water) and neutral formulations.

Health and medical applications include it as a material in hip replacements and birth control pills.

It is used as a dosimeter for radiation protection and therapy applications for its optically stimulated luminescence properties.

Aluminium oxide is an electrical insulator used as a substrate (silicon on sapphire) for integrated circuits but also as a tunnel barrier for the fabrication of superconducting devices such as single electron transistors and superconducting quantum interference devices (SQUIDs).

Aluminum oxide being a dielectric with relatively large band gap is used as an insulating barrier in capacitors.

Insulation for high-temperature furnaces is often manufactured from aluminium oxide. Sometimes the insulation has varying percentages of silica depending on the temperature rating of the material. The insulation can be made in blanket, board, brick and loose fiber forms for various application requirements.

Small pieces of aluminium oxide are often used as boiling chips in chemistry.

It is also used to make spark plug insulators.

Using a plasma spray process and mixed with titania, it is coated onto the braking surface of some bicycle rims to provide abrasion and wear resistance.

Most ceramic eyes on fishing rods are circular rings made from aluminium oxide.

Fluoroantimonic Acid

Fluoroantimonic acid is considered the strongest liquid superacid, and depending on who we talk to, may be considered the strongest superacid. It is an inorganic compound with the chemical formula H_2FSbF_6 (also written $H_2F[SbF_6]$, $2HF \cdot SbF_5$, or simply HF-SbF_5). It is 2×1019 (20 quintillion) times stronger than 100% sulfuric acid. Fluoroantimonic acid has a H0 (Hammett acidity function) value of -31.3.

The reaction to produce fluoroantimonic acid is:

$$SbF_5 + 2HF \rightarrow SbF_6^- + H_2F^+$$

The acid is often said to contain "naked protons", but the "free" protons are, in fact, always bonded to hydrogen fluoride molecules to make the fluoronium cations similar to the hydronium cation in aqueous solution. It is the fluoronium ion that accounts for fluoroantimonic acid's extreme acidity. Fluoroantimonic acid is 10 (10 quadrillion) times stronger than 100% sulfuric acid.

Structure and Properties of Fluoroantimonic Acid

Fluoroantimonic acid is a mixture of hydrogen fluoride and antimony pentafluoride in various ratios. The 1:1 combination forms the strongest known superacid, which has been demonstrated to protonate even hydrocarbons to afford carbocations and H_2.

Two related products have been crystallised from HF-SbF mixtures, and both have been analyzed by single crystal X-ray crystallography. These salts have the formulas $(H_2F^+)(Sb_2F_{11}^-)$ and $(H_3F_2^+)(Sb_2F_{11}^-)$. In both salts, the anion is $Sb_2F_{11}^-$. As mentioned above, SbF_6^- is weakly basic; the larger anion $Sb_2F_{11}^-$ is expected to be still weaker.

HF-SbF_5 is extremely corrosive, toxic, and moisture sensitive. Like most strong acids, fluoroantimonic acid can react violently with water, owing to the exothermic hydration. Consequently, it cannot be used in aqueous solution, only in hydrofluoric solution.

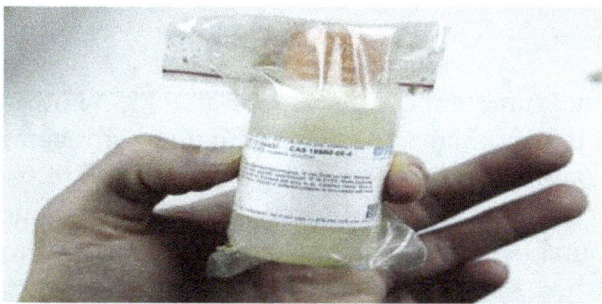

Applications of Fluoroantimonic Acid

Materials compatible with fluoroantimonic acid as a solvent include SO_2ClF, and sulfur dioxide; some chlorofluorocarbons have also been used. Containers for HF-SbF_5 are made of PTFE. This extraordinarily strong acid protonates nearly all organic compounds. In 1967, Bickel and Hogeveen showed that 2HF·SbF5 will remove H_2 from isobutane and methane from neopentane to form carbenium ions:

$$(CH_3)_3CH + H+ \rightarrow (CH_3)_3C+ + H_2$$

$$(CH_3)_4C + H+ \rightarrow (CH_3)_3C+ + CH_4$$

Digallane

Digallane is a chemical compound of gallium and hydrogen with the formula Ga2H6. The eventual preparation of the pure compound, reported in 1989, was hailed by N N Greenwood as a "tour de force". Historically digallane had been reported as early as 1941 by E. Wiberg, however this was not verified by later work by N.N Greenwood and others.

Preparation

A two stage approach proved to be the key to successful synthesis of pure digallane. Firstly the dimeric monochlorogallane, $(H_2GaCl)_2$ (containing bridging chlorine atoms this can be formulated as $(H_2Ga(m\text{-}Cl))_2)$ was prepared via the hydrogenation of gallium trichloride, $GaCl_3$ with Me_3SiH .This was followed by a further reduction with $LiGaH_4$, solvent free, at $-23°C$ to produce digallane, Ga_2H_6, in low yield.

$$Ga_2Cl_6 + 2\,Me_3SiH \longrightarrow (H_2GaCl)_2 + 2\,Me_3SiCl$$

$$(H_2GaCl)_2 + LiGaH_4 \longrightarrow Ga_2H_6 + LiCl$$

Digallane is volatile and condenses at $-50°$ C to give a white solid.

Structure and Bonding

Electron diffraction measurements of the vapour at $255°K$ show the structure of digallane to be similar to that of diborane with 2 bridging hydrogen atoms. The terminal Ga-H bond length is 152pm, the Ga-H bridging is 171 pm and the Ga-H-Ga angle is $98°$. The Ga-Ga distance is 258pm. The NMR spectrum of a digallane solution in toluene shows to peaks attributable to terminal and bridging hydrogen atoms.

The solid appears to be polymeric, and on the basis of its vibrational spectrum is possibly tetrameric (i.e. GaH3)4. There are terminal, non-bridging, hydrogen atoms which contrasts with α-alane, a high melting, relatively stable polymeric form of aluminium hydride, where all hydrogen atoms are equivalent and the aluminium atoms are 6 co-ordinate.

Reactions

Digallane decomposes at ambient temperatures:

$$2\,GaH_3 \longrightarrow 2\,Ga + 3\,H_2$$

The reactions with Lewis bases are similar to those of diborane. With trimethylamine 1:1 and 2:1 adducts are formed i.e. $NMe_3.GaH_3$ $(NMe_3)_2.GaH_3$. With phosphine a 1:1 adduct $PH_3.GaH_3$. The trimethylamine adducts have been known for some time. They were prepared by the reaction of $LiGaH_4$ with trimethylammonium chloride, Me_3NHCl, the 2:1 adduct is formed at low temperatures:

$$LiGaH_4 + Me_3NHCl \longrightarrow LiCl + H_2 + Me_3N.GaH_3$$

The structure of $Me_3N.GaH_3$ was reexamined and it has been confirmed to be monomeric in both the gas and solid phases. In this it contrasts with the corresponding alane adduct, NMe3.AlH3 which in the solid is dimeric with bridging hydrogen atoms. A range of other 1:1 adducts have been prepared and their stabilities determined.

References

- Haynes, William M., ed. (2011). CRC Handbook of Chemistry and Physics (92nd ed.). Boca Raton, FL: CRC Press. p. 4.45. ISBN 1439855110

- What-are-inorganic-compounds-definition-characteristics-examples: study.com, Retrieved 16 April 2018

- Aluminum chloride Archived 2014-05-05 at the Wayback Machine.. Chemister.ru (2007-03-19). Retrieved on 2017-03-17

- For isoelectric point: Gayer, K. H.; Thompson, L. C.; Zajicek, O. T. (September 1958). "The solubility of aluminum hydroxide in acidic and basic media at 25 ?c". Canadian Journal of Chemistry. 36 (9): 1268–1271. doi:10.1139/v58-184. ISSN 0008-4042. Retrieved 2017-07-01

- 2-4-inorganic-compounds-essential-to-human-functioning: library.open.oregonstate.edu, Retrieved 15 March 2018

- Washington, Neena (2 August 1991). Antacids and Anti Reflux Agents. Boca Raton, FL: CRC Press. p. 10. ISBN 0-8493-5444-7

- Perigrin, Tom. «Barium Chlorate». GeoCities. Archived from the original on 2007-10-30. Retrieved 2007-02-22

- Physical-characteristics-aluminum-oxide-8345240: sciencing.com, Retrieved 10 July 2018

- Hudson, L. Keith; Misra, Chanakya; Perrotta, Anthony J.; Wefers, Karl and Williams, F. S. (2002) «Aluminum Oxide» in Ullmann's Encyclopedia of Industrial Chemistry, Wiley-VCH, Weinheim. doi:10.1002/14356007.a01_557

3

Inorganic Reactions

Inorganic chemical reactions generally fall into the broad categories of combination, decomposition, single displacement and double displacement reactions. This chapter closely examines these crucial inorganic reactions.

Combination Reactions

Combination reactions describe a reaction like this:

$$A + B \rightarrow C$$

in which two or more reactants become one product (are combined). The problem with this term is that it doesn't give you much chemical insight because there are many different types of reactions that follow this pattern. So we'll break it into groups that reflect what's actually happening a little better.

Combination of Elements to Make Ionic Substances

In this category, an elemental metal and an elemental non-metal react to make an ionic substance that is neutral and has each ion in its correct charge state or valence. For instance,

$$2Na(s) + Cl_2(g) \rightarrow 2NaCl(s)$$

$$2Mg(s) + O_2(g) \rightarrow 2MgO(s)$$

$$2Al(s) + 3O_2(g) \rightarrow Al_2O_3(s)$$

If the metal is a transition metal, it will be much harder to predict the correct charge on the metal ion in the ionic compound.

Often, an elemental metal and non-metal "want" to make an ionic compound, because this is a more stable state (think about a heavy ball on a table: it can easily roll to the ground, where it has less potential energy, so the table isn't a stable state; if the heavy ball is in a small hole in the ground, it can't easily move, and if it did, it would have more potential energy, so the hole is a stable, low energy state). However, that doesn't necessarily mean the reaction will just happen on its own. That depends on how easily the reaction can happen (think about a place you want to go, but don't go because traveling there is very inconvenient).

A ball being held above a ledge has more potential energy so it is in a less stable state than if it were at the bottom of the ledge.

For instance, the alkali metals and the halogens react pretty easily, so they will often react even without a "push." Oxygen is very reactive, which is why things burn, but you have to light them on fire to get them started. This is good, because otherwise we would burn in air at room temperature! Many of these elemental combination reactions might need a high temperature to get started, even if they want to happen. It won't be hard to remember that alkalis, alkaline earth metals and halogens react easily, because this is why they are very hard to find in elemental form! Oxygen and nitrogen are very abundant in elemental form because it is hard for them to react even if they want to. Nitrogen in particular reacts only with lithium metal and a few complicated compounds at room temperature, although it will react with many other elements at high temperatures. Most metals aren't found in elemental form in nature (except for ones that don't want to react, like gold), but if you find them in elemental form in your house, then probably they don't react easily.

Combination of Elemental Non-metals into Covalent Compounds

These reactions involve elemental forms of elements like H, C, N, O, Cl, S, P, etc. It will often be hard to predict the product because these elements can often combine in different ratios (this is where the law of multiple proportions comes from!). You can always expect that H will have a valence of 1, and O will usually have a valence of 2. Many of these reactions will happen quickly if you get them started with a little heat, especially if oxygen or a halogen is involved. Otherwise, they might happen very slowly or not at all except under special circumstances that we will talk more about later. Some examples:

$$C(s) + O_2(g) \rightarrow CO_2(g) (fast, once\,lit)$$

$$N_2(g) + 3H_2(g) \rightarrow 2NH_3(g) (usually\,very\,slow)$$

Basic Anhydrides

Basic anhydrides are compounds that turn into a base (a hydroxide salt) when you add water. They are metal oxides. Here's an example:

$$CaO(s) + H_2O(l) \rightarrow Ca(OH)_2(aq)$$

If the metal is an alkali or alkaline earth, the reaction probably happens quickly and produces a lot of heat. If the metal is a transition metal, the reaction might not happen so easily or at all.

Acid Anhydrides

Acid anhydrides are compounds that turn into an acid when you add water. They are non-metal oxides. These are a little more complicated than basic anhydrides, so don't worry too much about them right now. Here's an example:

$$SO_3(g) + H_2O(l) \rightarrow H_2SO_4(aq)$$

Decomposition Reactions

A decomposition reaction is a type of chemical reaction in which a single compound breaks down into two or more elements or new compounds. These reactions often involve an energy source such as heat, light, or electricity that breaks apart the bonds of compounds.

It can be represented by the general equation:

$$AB \rightarrow A + B$$

In this equation, AB represents the reactant that begins the reaction, and A and B represent the products of the reaction. The arrow shows the direction in which the reaction occurs.

In a combination reaction, a substance is formed as a result of chemical combination, while in a decomposition reaction, the substance breaks into new substances.

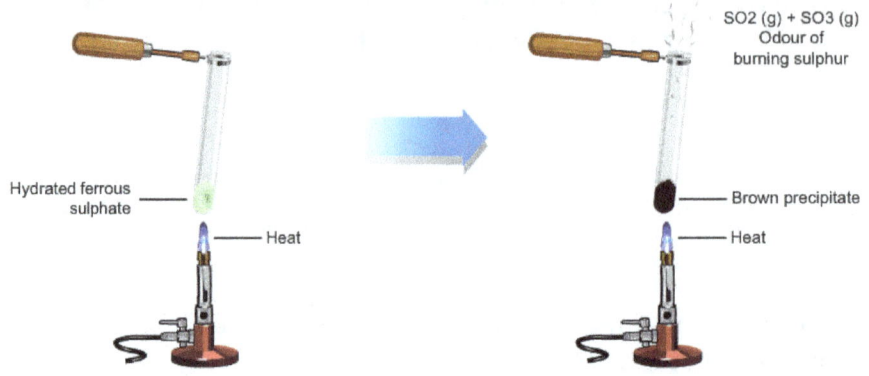

Thermal Decomposition

For example: The digestion of food in our body is accompanied by a number of decomposition reactions. The major constituents of our food such as carbohydrates, fats, proteins, etc., decompose to form a number of simpler substances. These substances further react, releasing large amounts of energy, which keeps our body working.

Types of Decomposition Reactions

Decomposition reactions can be classified into three types:

- Thermal decomposition reaction
- Electrolytic decomposition reaction
- Photo decomposition reaction

Thermal decomposition is a chemical reaction where a single substance breaks into two or more simple substances when heated. The reaction is usually endothermic because heat is required to break the bonds present in the substance.

Examples:

Photo decomposition is a chemical reaction in which a substance is broken down into simple substances by exposure to light (photons).

Thermal decomposition reaction (Thermolysis)

- Decomposition of calcium carbonate: Calcium carbonate (lime stone) decomposes into calcium oxide (quick lime) and carbon dioxide when heated. Quick lime is the major constituent of cement.

$$CaCO_3(s) \xrightarrow{\text{Heat}} CaO(s) + CO_2(g)$$

- Decomposition of potassium chlorate: When heated strongly, potassium chlorate decomposes into potassium chloride and oxygen. This reaction is used for the preparation of oxygen.

$$2\,KClO_3(S) \xrightarrow{\text{Heat}} 2\,KCl(S) + 3\,O_2(g)$$

If the decomposition is carried out in the presence of manganese dioxide (MnO_2), it takes place at a lower temperature. In this case, MnO_2 is used as a catalyst.

- Decomposition of ferric hydroxide: Ferric hydroxide undergoes decomposition in the presence of heat, giving ferric oxide and water molecules.

$$2\,Fe(OH)_3 \xrightarrow{\text{Heat}} Fe_2O_3 + 3\,H_2O$$

Decomposition of hydrated oxalic acid:

Hydrated oxalic acid $(H_2C_2O_4.2H_2O)$ decomposes into oxalic acid and water when heated.

$$H_2C_2O_4.2H_2O \xrightarrow{\text{Heat}} H_2C_2O_4 + 2\,H_2O$$

Electrolytic Decomposition Reaction (Electrolysis)

Electrolytic decomposition may result when electric current is passed through an aqueous solution of a compound. A good example is the electrolysis of water.

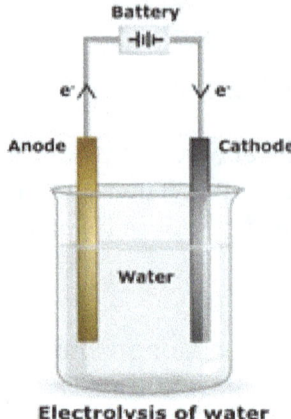

Electrolysis of water

$$2\,H_2O(l) \xrightarrow{\text{Electric current}} 2\,H_2(g) + O_2(g)$$

$$2\,NaCl \xrightarrow{\text{Electricity}} 2\,Na + Cl_2$$

- Electrolysis of water: Electrolysis of water is the decomposition of water into hydrogen and oxygen due to the passage of electric current through it.

- Decomposition of sodium chloride: On passing electricity through molten sodium chloride, it decomposes into sodium and chlorine.

Photo Decomposition Reaction (Photolysis)

- Decomposition of silver chloride: Place a small quantity of silver chloride (AgCl) taken in a watch glass under sunlight for some time. The crystals slowly acquire a grey colour. On analysis, it is found that the sunlight has caused decomposition of silver chloride into silver and chlorine.

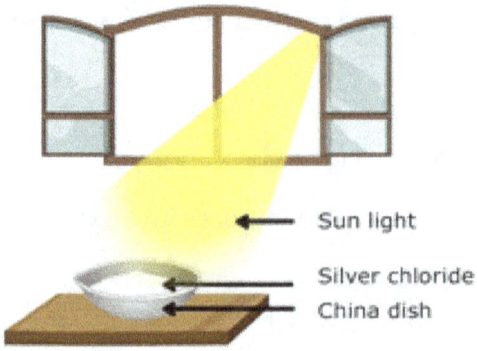

Decomposition of silver chloride

$$2\,\text{Ag}\,\text{cl}\,(\text{s})\xrightarrow{\text{Sunlight}}2\,\text{Ag}(\underset{\text{Grey}}{\text{s}})+\text{Cl}_2(\text{g})$$
white

Silver bromide also decomposes in the same way.

$$2\,AgBr(s)\xrightarrow{\;sunlight\;}2\,Ag(s)+Br(g)$$
LightYellow   Grey

- Decomposition of hydrogen peroxide: In the presence of light, hydrogen peroxide decomposes into water and oxygen

$$2\,\text{H}_2\text{O}_2(\text{l})\xrightarrow{\;\text{Light}\;}2\,\text{H}_2\text{O}(\text{l})+\text{O}_2(\text{g})$$

Single Displacement Reactions

A single replacement reaction, sometimes called a single displacement reaction, is a reaction in which one element is substituted for another element in a compound. The starting materials are always pure elements, such as a pure zinc metal or hydrogen gas, plus an aqueous compound. When a replacement reaction occurs, a new aqueous compound and a different pure element will be generated as products. The general pattern of a single replacement reaction is shown below.

$$\text{AB(aq)}+\underset{\substack{\downarrow\\ Pure}}{\text{C}}\longrightarrow\underset{\substack{\downarrow\\ elements}}{\text{A}}+\text{CB(aq)}$$

Determining the Products of Single Replacement Reactions

If we are trying to figure out whether a single displacement reaction will occur, there are two main questions we need to answer

Two Elements that Might Swap Places in our Proposed Reaction

In general, elements that form anions can replace the anion in a compound, and elements that form cations can replace the cation in a compound. The following guidelines can be used to determine what kind of ions a given element might form.

- Metals will usually form cations. This includes groups 1 and 2, some of group 13 and 14 elements, and the transition metals.

- The common non-metals in single replacement reactions are the group 17 elements, which generally form anions with a 1- charge.

- Hydrogen usually forms the cation H^+ in a single replacement reaction.

H, start superscript, plus, end superscript in a single replacement reaction.

In our reaction with copper metal and aqueous silver(I) nitrate, the copper metal will likely react to form copper cations because it is a transition metal. The copper cations can replace the silver cations in the compound $AgNO_3(aq)$ to form a new compound.

New Compound that Will Form as a Product

Once we know what element might be replaced in our ionic compound, we can predict the products that might be formed. In this example, the silver atoms in $AgNO_3(aq)$ can be replaced by copper to form $Cu(NO_3)_2$ aq. In the process, elemental silver, $Ag(s)$ would also form as a product. We can write out the full—and balanced!—reaction as follows:

$$2\,AgNO_3(aq) + Cu(s) \longrightarrow Cu(NO_3)_2(aq) + 2\,Ag(s)$$

Does this match our observations? It turns out that aqueous solutions of $Cu(NO_3)_2$ are blue-green, which explains the solution's color change. The grey fuzz growing on the copper would be from silver metal precipitating out on the surface of the wire.

Predicting if a single Replacement Reaction Will Occur

Once we know which elements might get swapped in our single displacement reaction, we can predict whether the reaction will occur based on knowledge of the relative reactivities of the two elements—elements C and A in the generic pattern above, or copper and silver in our example reaction. If element C is more reactive than element A, then C will replace A in a compound. If element C is less reactive than element A, then there will be no reaction.

The reactivity series—also called the activity series—ranks elements in order of their reactivity for certain types of reactions, including single replacement reactions. The more reactive elements will replace the less reactive elements in the reactivity series, but not the other way around. There are separate rankings for elements that form cations and elements that form anions.

For elements that tend to gain electrons to form anions, the order of reactivity from most reactive to least reactive goes as follows:

$$\textit{Most reactive} \quad F_2 > Cl_2 > Br_2 > I_2 \quad \textit{Least reactive}$$

For these elements, you can also look at their arrangement on the periodic table—group 17—to remember the order of reactivity. The higher the element's position in the column, the more reactive it will be. Based on this activity series, we would predict that Br_2 would replace I_2 in a single replacement reaction, but Br_2 would not react with a compound containing fluoride ions.

For the cation-forming elements, the reactivity series is longer, and the trends are not as straightforward. You can see an example of the reactivity series for cations below.

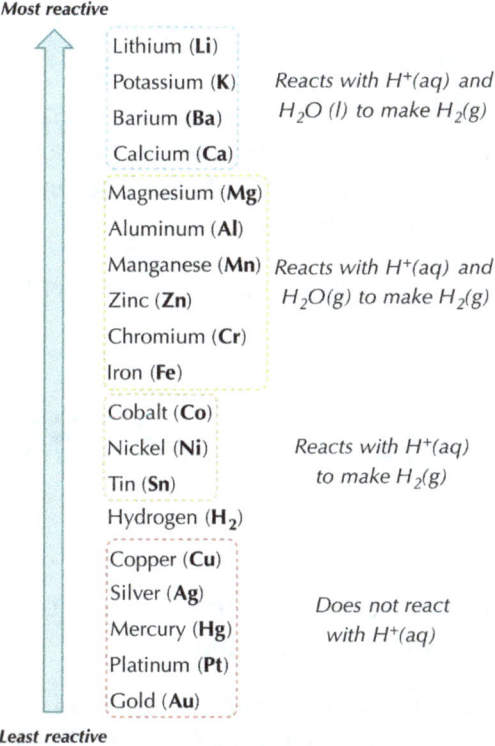

Reactivity is pretty complicated! After all, there are lots of different kinds of reactions, so what kind of reactivity are we really ranking here? Some properties that are taken into account in the reactivity series include reactivity with water and acids, as well as how readily an element loses electrons to form cations. As a result of the different ways reactivity can be defined, however, you might see some elements ranked in a different order depending on your teacher or textbook.

The process of using the reactivity series is the same for both cations and anions:

More reactive elements will replace less reactive elements in a compound.

Single Displacement Reaction Examples

The reaction between zinc metal and hydrochloric acid to produce zinc chloride and hydrogen gas is an example of a single displacement reaction:

$$Zn(s) \ + \ 2HCl(aq) \ \rightarrow \ ZnCl_2(aq) \ + \ H_2(g)$$

Another example is the displacement of iron from an iron(II) oxide solution using coke as a carbon source:

$$2\,Fe_2O_3(s) + 3C(s) \longrightarrow Fe(s) + CO_2(g)$$

Double Displacement Reactions

Double displacement reactions may be defined as the chemical reactions in which one component each of both the reacting molecules is exchanged to form the products. During this reaction, the cations and anions of two different compounds switch places, forming two entirely different compounds.

The general equation which represents a double displacement reaction can be written as:

Double displacement reactions generally take place in aqueous solutions in which the ions precipitate and there is an exchange of ions.

For example, on mixing a solution of barium chloride with sodium sulphate, a white precipitate of barium sulphate is immediately formed. These reactions are ionic in nature. The reactants changes into ions when dissolved in water and there is an exchange of ions in solution. This results in the formation of product molecule.

$$Ba^{2+}(aq) + 2Cl^-(aq) + 2Na^+(aq) + SO_4^{2-}(aq) \longrightarrow Ba^{2+}(aq) + SO_4^{2-}(aq)2Na(aq) + 2Cl^-(aqa)$$

Double Displacement Reactions Can be Further Classified as Neutralization, Precipitation and Gas Formation Reactions

Neutralization reactions are a specific kind of double displacement reaction. An acid-base reaction occurs, when an acid reacts with equal quantity of base. The acid base reaction results in the formation of salt (neutral in nature) and water.

$$HCl(aq) + NaOH(aq) \longrightarrow NaCl(aq) + H_2O(l)$$

Precipitation is the formation of a solid in a solution or inside another solid during a chemical reaction. This process usually takes place when the concentration of dissolved ions in the solution exceeds the solubility product.

$$AgNO_3(aq) + NaCl(aq) \longrightarrow AgCl(s) + NaNO_3(aq)$$

$$BaCl_2(aq) + CuSO_4(aq) \longrightarrow BaSO_4(s) + CuCl(aq)$$

$$PbNO_3(aq) + Na_2SO_4(aq) \longrightarrow PbSO_4(s) + 2NaNO_3(aq)$$

$$CuSO_4(aq) + H_2S(g) \longrightarrow CuS(s) + H_2SO_4(aq)$$

$$Pb(NO_3)_2(aq) + 2\,NaI(aq) \longrightarrow PbI_2(s) + 2\,NaNO_3(aq)$$

$$CoCl_2(aq) + Na_2CO_3(aq) \longrightarrow CoCO_3(aq) + 2\,NaCl(aq)$$

$$Al_2(SO_4)_3(aq) + 3\,Ca(OH)_2(aq) \longrightarrow 2\,Al(OH)_3(aq) + 3\,CaSO_4(s)$$

$$Pb(CH_3COO)_3(aq) + 2\,HCl(aq) \longrightarrow PbCl_2(s) + CH_3COOH(aq)$$

A double displacement reaction should also occur if an insoluble gas is formed. Gases such as HCl and NH_3 are soluble in water, but some other gases, such as $H\,S$, are not soluble in water.

$$ZnS(s) + 2\,HCl(aq) \longrightarrow ZnCl_2(aq) \longrightarrow ZnCl_2(aq) + H_2S(g)$$

- Neutralization Reactions

  On mixing an aqueous solution of hydrochloric acid with an aqueous solution of sodium hydroxide, sodium chloride and water are formed.

- Precipitation Reactions

  On mixing aqueous solutions of silver nitrate and sodium chloride, a white curdy precipitate of silver chloride is formed.

  On mixing an aqueous solution of barium chloride with that of copper sulphate, a white precipitate of barium sulphate is formed.

  On mixing an aqueous solution of lead nitrate with sodium sulphate, a white precipitate of lead sulphate is formed.

  On passing hydrogen sulphide gas through copper sulphate solution, a black precipitate of copper sulphide is formed.

  On adding a solution of lead nitrate to sodium iodide solution, a yellow precipitate of lead iodide is formed.

  Cobalt(II) chloride reacts with sodium carbonate to form pink/red coloured precipitate of cobalt(II) carbonate and sodium chloride.

  On adding aluminium sulphate solution to calcium chloride solution, a precipitate of calcium sulphate is formed.

  Lead acetate solution is treated with dilute hydrochloric acid to form lead chloride and acetic acid solution.

- Gas Formation Reactions

    Many sulfide salts will react with acids to form gaseous hydrogen sulphide.

    $$CaCO_3(s) + 2HCl(aq) \longrightarrow CaCl_2(aq) + H_2CO_3(aq)$$

    $$H_2CO_3(aq) \longrightarrow CO_2(g) + H_2O(l)$$

We may conclude that in double displacement reactions, two compound exchange ions or elements to form a new compound. The product obtained is either a gas or solid which can be separated from the reaction mixture, or a stable covalent compound, often water.

Water-gas Shift Reactions

Water gas shift reactions are carried out at conditions of high and low temperature. For the water gas shift reaction, the equilibrium favors the reaction at lower temperature. However, copper based catalyst, active at low temperature, is highly sensitive to residual sulfur or chloride compounds which have passed through the pre-purification steps. Hence, initially the reaction is carried out over catalyst composed of Fe_2O_3 and Cr_2O_3 which is relatively inexpensive, resistant to sintering and tolerant to sulfur and chloride poisoning. However, at low temperature Fe_3O_4 is not sufficiently active hence, the reaction has to be carried out at higher temperature. The high temperature catalyst adsorbs the residual sulfur or chloride compounds thereby, protecting the low temperature copper based shift catalyst from exposure to these compounds. The high temperature process is followed by shift reaction at thermodynamically favorable lower temperature of 200°C over copper based catalyst.

High Temperature Water Gas Shift Reaction

Gas from the secondary reformer contain 10-13 % CO and is further processed to increase the H_2 concentration. The high temperature water gas shift reaction lowers the CO concentration to about 2-3 % in a fixed bed reactor.

$$CO + H_2O = CO_2 + H_2$$

Process is operated at 350-500°C and 20 -30 atm. Since equilibrium favors the WGS at lower temperature, the feed gas is cooled to 350-400°C before entering the reactor.

Catalysts

A typical catalyst is composed of 90 % Fe_3O_4 and 10 % Fe_2O_3. Fe_3O_4 acts as the active metal while Fe_2O_3 acts as a stabilizer to minimize sintering of the active iron oxides. The catalysts are prepared by precipitation.

Kinetics

The rate expression from Langmuir Hinshelwood model proposed is given as :

$$r = \frac{kK_{CO}K_{H_2O}\left(P_{CO}P_{H_2O}P_{CO_2}P_{H_2}\right)^2}{\left(1 + K_{CO}P_{CO} + K_{H_2O}P_{H_2O} + K_{CO_2}P_{CO_2} + K_{H_2}P_{H_2}\right)^2}$$

k is the rate constant and K_i's the respective adsorption constants. The 'i' represents the chemical species.

Activation energy for the water gas shift reaction on iron oxide/chromia is in the range of 122 kJ/mol. Reaction over small pellets (5 x 4 mm) operates with an effectiveness factor of unity below 370°C and 31 atm.

Deactivation

The catalyst is deactivated by adsorption of residual S or Cl containing compounds that escape the initial purification steps.

Low Temperature Water Gas Shift

The CO concentration of the existing gas from high temperature WGS is 2-3 % and further reduced to below 0.2 % by low temperature water gas shift reaction. The WGS reaction as favored at low temperature, operating temperature is maintained at 200°C, the minimum temperature at which steam is not condensed and operating pressure at 10-30 atm.

Catalysts

A typical catalyst is composed of CuO/ ZnO/Fe_2O_3 and prepared by co-precipitation method. The catalyst is highly selective for shift reaction with low activity for meth-anation reaction. CuO is the catalytically active metal and ZnO minimize sintering of Cu. Cu is highly sensitive to poisoning by S or Cl compounds even at very low level of 1 ppm. ZnO & Fe_2O_3 also scavenge the S or Cl compounds protecting Cu which is highly sensitive to poisoning by these compounds. Catalyst is reduced very carefully by H_2 and the bed temperature is never allowed to rise above 230°C to avoid sintering of Cu. The high temperature shift catalyst also protects the low temperature catalyst from these poisons.

Kinetics

Kinetic model is based on surface redox mechanism involving dissociation of water to OH radicals which further dissociate to atomic oxygen. The rate expression as proposed by Ovensen is

$$r = A\exp\left(\frac{-E_a}{RT}\right)p_{CO}^{m}p_{H_2O}^{n}p_{\infty_2}^{x}p_{H_2}^{y}\left(1-\beta\right) \quad \beta = \frac{1}{K_{WGS}}\frac{P_{CO_2}P_{H_2}}{P_{CO}P_{H_2O}}$$

A= pre-exponential factor; E_a= activation energies; m, n, x, y = respective reaction orders; β defined as approach to equilibrium and K_{WGS} is equilibrium constant for WGS reaction.

Final CO/CO$_2$ Removal by Methanation

After low temperature WGS, the CO content in hydrogen stream is 0.2-0.5 % and must be reduced to 5 ppm level for using the H_2 for ammonia synthesis. The CO is further reacted with hydrogen and converted to methane. The catalyst is typically Ni/Al$_2$O$_3$. MgO is sometimes used as a promoter to minimize sintering of Ni. The process is carried out at 300 -350°C and at 30 atm pressure. Care must be taken to avoid formation of poisonous Ni(CO)$_6$ which occurs below 200°C at partial pressure of CO greater than ~ 0.2 atm.

4

Bioinorganic Chemistry

Bioinorganic chemistry delves into the role of metals in the field of biology. It involves the study of natural phenomena like the behavior of metalloproteins and artificially introduced metals in medicine and toxicology. An elaborate study of the varied principles of bioinorganic chemistry has been provided in this chapter.

Bio-inorganic Chemistry is an interdisciplinary area composed of mainly Biochemistry and Inorganic Chemistry. Bioinorganic chemistry is the understanding of the influence or role of inorganic materials (mainly metal ions) to the biological processes, *e . g .* electron transport, ion transport, mineralization of inorganic materials, mutation, inorganic species in medicinal therapy and diagnosis, etc. In short Bioinorganic Chemistry is the understanding of the influence or role of inorganic materials (mainly metal ions) to the biological processes.

The basic unit of living organism is cell. Prokaryotic cells are simple and found mainly in bacteria and bacteria like organisms. Large and much complex cells found in animals and plants are known as eukaryotic cells.

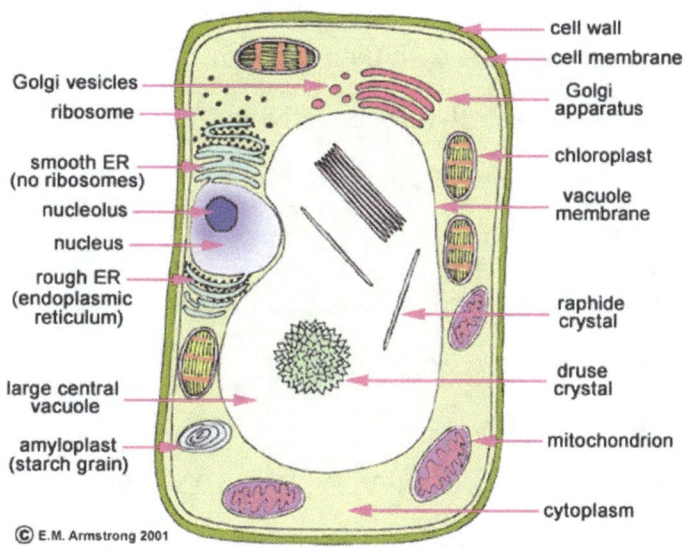

Figure: A eukaryotic cell representation

A prokaryotic cell consists of an enclosed aqueous phase called cytoplasm and it contains the DNA and most of the other materials used for the biological transformation reactions. The aqueous phase is surrounded by a single membrane or an additional intermediate aqueous phase between the outer most membrane and the inner membrane. This aqueous phase is known as periplasm.

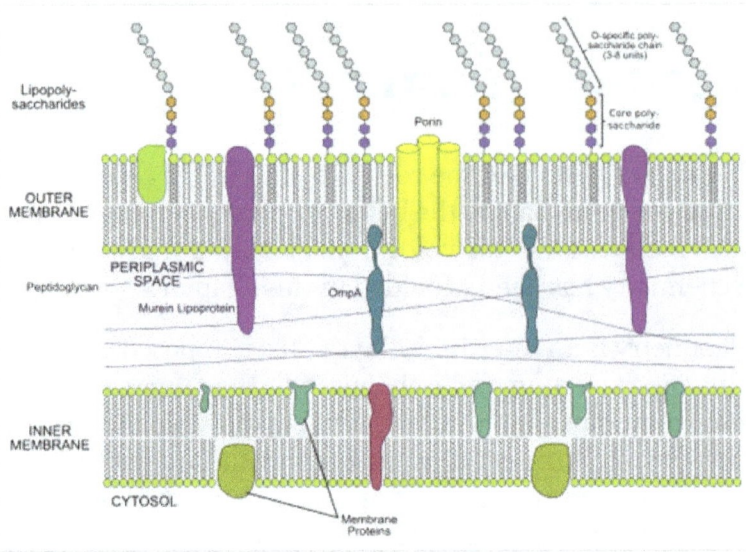

Figure: Showing periplasm

The cytoplasm of eukaryotic cell contains several sub compartments which are enclosed within lipid bilayers and known as organelles. These include nucleus (contains DNA), mitochondria (power house of the cell), chloroplast (photo cell), Golgi (vesicles containing proteins for export).

Some biological essential elements with their functions:

- Charge balance and electrolytic conductivity: Na, K, Cl

- Structure and templating: Ca, Zn, Si, S, Mo, Ni

- Signaling: Ca, B, NO

- Brønstead Acid-Base Buffering: P, Si, C

- Lewis Acid-Base Catalysis: Zn, Fe, Ni, Mn

- Electron Transfer: Fe, Cu,

- Group Transfer (e.g. CH_3, O, S): V, Fe, Co, Ni, Cu, Mo, W

- Redox Catalysis: V, Mn, Fe, Co, Ni, Cu, W, S, Se

- Energy Storage: H, P, S, Na, K, Fe

- Biomineralization: Ca, Mg, Fe, Si, Sr, Cu, P

- Energy generation: Ca, Mg

Essential Amino Acids

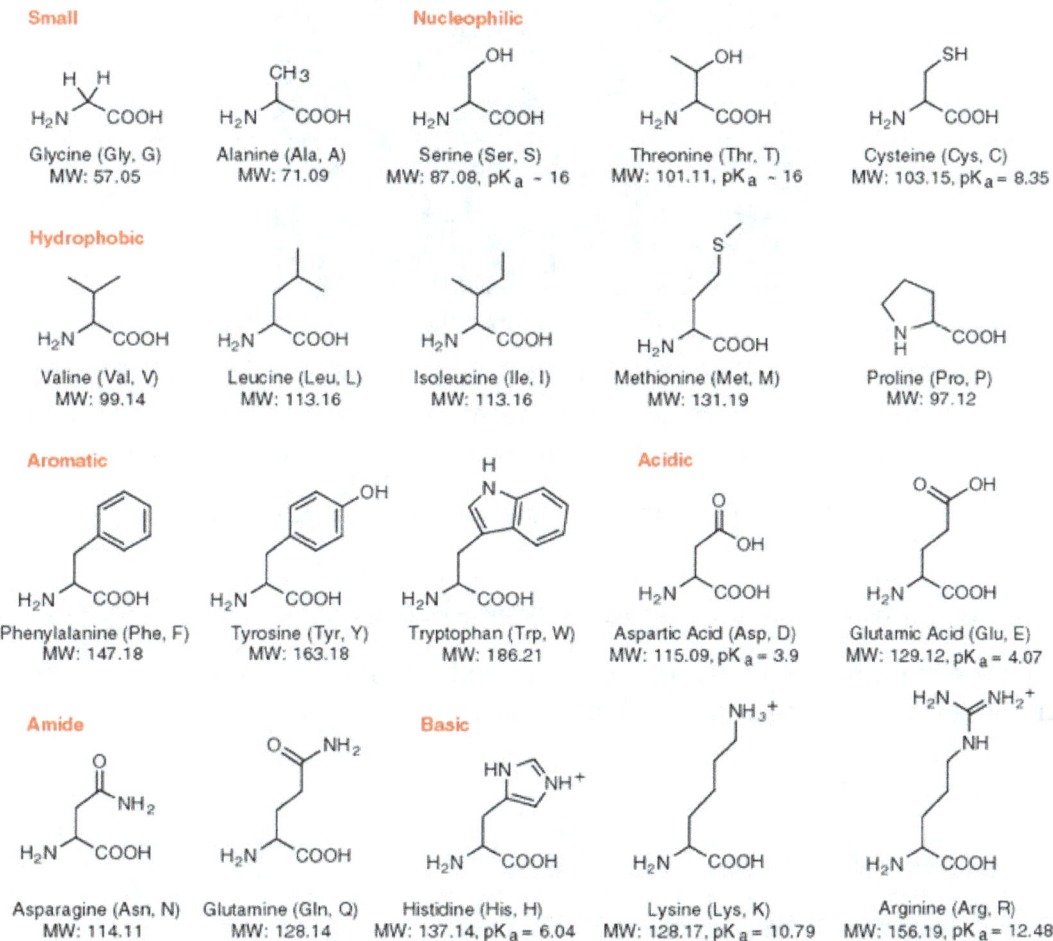

Small

Glycine (Gly, G)
MW: 57.05

Alanine (Ala, A)
MW: 71.09

Nucleophilic

Serine (Ser, S)
MW: 87.08, pK$_a$ ~ 16

Threonine (Thr, T)
MW: 101.11, pK$_a$ ~ 16

Cysteine (Cys, C)
MW: 103.15, pK$_a$ = 8.35

Hydrophobic

Valine (Val, V)
MW: 99.14

Leucine (Leu, L)
MW: 113.16

Isoleucine (Ile, I)
MW: 113.16

Methionine (Met, M)
MW: 131.19

Proline (Pro, P)
MW: 97.12

Aromatic

Phenylalanine (Phe, F)
MW: 147.18

Tyrosine (Tyr, Y)
MW: 163.18

Tryptophan (Trp, W)
MW: 186.21

Acidic

Aspartic Acid (Asp, D)
MW: 115.09, pK$_a$ = 3.9

Glutamic Acid (Glu, E)
MW: 129.12, pK$_a$ = 4.07

Amide

Asparagine (Asn, N)
MW: 114.11

Glutamine (Gln, Q)
MW: 128.14

Basic

Histidine (His, H)
MW: 137.14, pK$_a$ = 6.04

Lysine (Lys, K)
MW: 128.17, pK$_a$ = 10.79

Arginine (Arg, R)
MW: 156.19, pK$_a$ = 12.48

Peptide: Polymers of monomeric amino acids are called peptides and the linkage bond between two amino acids is called peptide linkage.

Peptide linkage

Enzyme: A biologically active compound containing one or more polypeptide units that are folded in a globular or fibrous form and catalyzes chemical reactions is called enzyme.

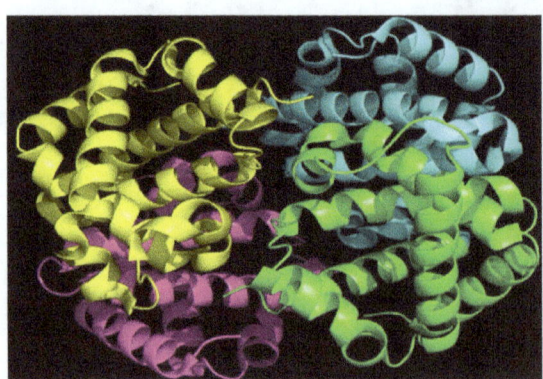

Figure: Showing folding of polypeptide chain in protein.

Apoenzyme: Many enzymes required an additional molecule to catalyze the particular chemical reaction. The small molecule is known as cofactor. It could be metal ion(s) or non-protein organic molecules. An enzyme without a cofactor is called apoenzyme.

Holoenzyme: An enzyme with a complete complement of cofactors is known as a holo-enzyme.

So it can be written; Holoenzyme = Apoenzyme + coenzyme

Metalloenzyme: Enzyme that contains metal ion(s) in its active site and metal ion(s) participate(s) in the biological transformation.

Enzyme Kinetics:

$$E + S \underset{K_{-1}}{\overset{K_1}{\rightleftharpoons}} E \bullet S \xrightarrow{K_2} P + E$$

E = enzyme, S = substrate, and P = Product

The substrate (S) binds reversibly to the enzyme (E) in the first step of an enzymatic reaction and forms $E \bullet S$. After that the product is formed with release of enzyme.

The rate of $E \bullet S$ formation is designated by k_1. The rates of decomposition of $E \bullet S$ are k_{-1} and k_2.

At the steady state, the rate of formation of $E \bullet S$ will be equal to the rate of decomposition of $E \bullet S$.

Hence, $k_1 [S][E_{free}] = k_{-1} \cdot [E \bullet S] + k_2 [E \bullet S]$, where $[E_{total}] = [E_{free}] + [E \bullet S]$

$k_1 \cdot [S] \cdot ([E_{total}] - [ES]) = k_{-1} \cdot [ES] + k_2 \cdot [ES]$

Solving for ES;

$$[ES] = \frac{k_1 \cdot [E_{total}][S]}{k_1 \cdot [S] + k_2 + k_{-1}} = \frac{[E_{total}][S]}{[S] + \frac{k_2 + k_{-1}}{k_1}}$$

The velocity of the enzyme reaction therefore is;

$$\text{Velocity} = k_2 \cdot [ES] = \frac{k_2 \cdot [E_{total}][S]}{[S] + \dfrac{k_2 + k_{-1}}{k_1}}$$

Finally, define V_{max} (the velocity at maximal concentrations of substrate) as k_2 times E_{total}, and K_M, the Michaelis-Menten constant, as $(k_2 + k_{-1})/k_1$. Substituting;

$$\text{Velocity} = V = \frac{V_{max}[S]}{[S] + K_M}$$

$$\text{Hence,} \quad \frac{1}{V} = \frac{K_M}{V_{max}[S]} + \frac{1}{V_{max}}$$

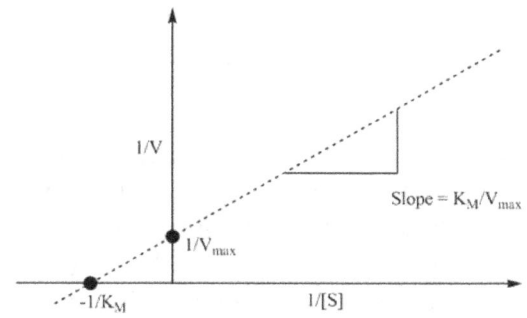

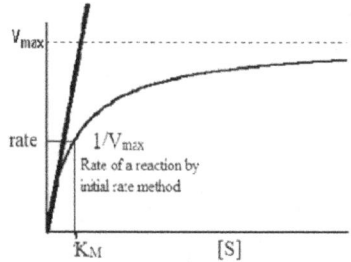

Some important notes:

At low values of [S], the initial velocity of a enzymatic reaction, V, rises almost linearly with increasing of substrate concentration, [S].

But as the concentration of [S] increases, the gains in V level off (forming a rectangular hyperbola).

The asymptote represents the maximum velocity of the reaction, designated V_{max}.

The substrate concentration that produces a V that is one-half of V_{max} is designated the Michaelis-Menten constant, K_M.

K_M is (roughly) an inverse measure of the affinity or strength of binding between the enzyme and its substrate, i.e. E•S formation.

The lower the K_M, the greater the affinity (so the lower the concentration of substrate needed to achieve a given rate).

Types of Inorganic Elements in Biology

Alkali and Alkaline Earth Metals

The elements in group one of the periodic table (with the exception of hydrogen - see below) are known as the *alkali metals* because they form alkaline solutions when they react with water. This group includes the elements *lithium, sodium, potassium, rubidium, caesium* and *francium*. Each of these elements has just one valence electron, which means that they form only weak metallic bonds. As a result, they are relatively soft and have low melting points.

The single valence electron is easily lost, making these metals highly reactive. They react vigorously with both air and water - when sodium comes into contact with water, for example, it reacts violently to form sodium hydroxide and hydrogen. The heat of the reaction actually ignites the hydrogen! Alkali metals also readily combine with the elements of group seventeen (*chlorine, fluorine, bromine* etc.) to form stable *ionic compounds* like sodium chloride.

The alkali metals (highlighted) occupy group one in the periodic table

Group two of the periodic table comprises the elements *beryllium, magnesium, calcium, strontium, barium* and *radium*. The elements in this group, which are all shiny and silvery-white in appearance, are known as the *alkaline earth metals*. Like the alkali metals, they form alkaline solutions when they react with water. The term "earth" is historical; it was the generic name used by alchemists for the *oxides* of these elements (which at one time were thought to be elements in their own right).

All of the elements of group two have *two* electrons in their outer shell. Metallic bonds in the alkaline earth metals are thus stronger than for the alkali metals, resulting in higher melting points, but they are still quite reactive because the two outer electrons are easily lost. As a result, they are not found in nature in their elemental state.

All but one of the alkaline earth metals react with the halogens (*chlorine, fluorine* etc.) to form *ionic compounds* (beryllium chloride is the exception, because the bonding is covalent). All of the alkaline earth metals except beryllium and magnesium also react with water to produce hydrogen gas and their respective hydroxides (magnesium *will* react with steam, however). Essentially, the heavier the alkaline earth metal, the more vigorously it will react with water.

The alkaline earth metals (highlighted) occupy group two in the periodic table

Magnesium is the fifth most abundant element on earth, closely followed by calcium in eigth place - which is just as well, since both magnesium and calcium are vital to all living things, including human beings! Magnesium plays a part in a huge array of biochemical reactions; among other things, it is essential for healthy bones and teeth. Calcium makes up roughly two percent of our total body weight, with most of it residing in our teeth and bones.

Transition Metals

Elements that lose electrons easily, that are lustrous and malleable, and that are good conductors of heat and electricity are known as metals. Metal elements can be broken down into several categories, one of which is the category of transition metals.

A transition metal is defined as a metal with inner d or f orbitals being filled. Orbitals describe ways that electrons can be organized around a nucleus. There are four types of orbitals: s, p, d, and f.

The transition metals consist of the 40 elements located in columns 3-12 on the periodic table and the 28 elements comprising the lanthanide and actinide series. Elements in the lanthanide and actinide series are often considered to be inner transition metals. The transition metals are identifiable in the periodic table shown here in peach; the inner transition metals are dark pink and light pink in color. The d-block transition metals are in columns 3-12, often labeled 1B-10B in other versions of the periodic table. The f-block inner transition metals are in the two long rows below the periodic table. Inner transition metals are often just referred to as transition metals.

Properties of Transition Metals

Transition metals show similar properties by column and by row. In general, transition metals are lustrous, silvery, hard, and good conductors of heat and electricity. Properties between individual elements may vary greatly. For instance, mercury is a liquid at room temperature, whereas tungsten does not melt until 3,400 degrees Celsius.

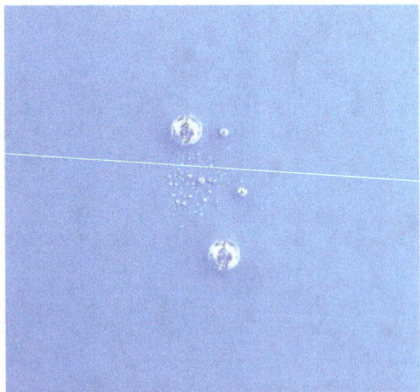

Mercury, often known as quicksilver, is a liquid at room temperature.
It is one of the oldest metals known to humans.

Some elements are extremely malleable, like gold and silver, while others, like cobalt, are more difficult to mold. Some metals, like copper, are very ductile and can be made into wires. Nearly all of the metals are good conductors of heat and electricity, but some are better than others. Copper and silver are amongst the best conductors.

Unlike elements from the rest of the periodic table, transition metals are comfortable losing different numbers of electrons. Whereas elements from column one can only ever form charges of +1, a single transition metal may be able to form variously charged ions. Iron, for example, can exist with a +2 or +3 charge.

Vanadium can exist with a charge between +2 and +5. The ability to have multiple types of cations is related to the organization of electrons in the d and f orbitals.

Chemical reactivity of the transition metals varies greatly. Some metals will react to form compounds, while others prefer to remain in their pure form. Many transition metals, like iron and titanium, will readily react with oxygen to form oxides. Other transition metals, like gold and platinum, are less reactive and can lie for thousands of years next to oxygen without reacting.

Famously unreactive gold is often found in its pure form.

Bioorganometallic Chemistry

Bioorganometallic chemistry refers to the study of organometallic compounds containing moieties of biological or medicinal interest, such as amino acids, enzymes, vitamins, DNA, and sugars. Due to the immense variety of organic compounds which have roles in biological processes, this results in a number of areas in which bioorganometallic compounds maybe able to play a role in. Currently the main fields of bioorganometallic research are in molecular recognition, enzymes and proteins, toxicology, therapy, and bioanalysis.

Bioorganometallics and Toxicology

There have been many toxicological and environmental problems associated with organometallic compounds. Famous historical cases of organometallic poisoning include organomercury poisoning in Japan due to water pollution from factories, the use of organoarsenic compounds as toxic gases during World War I,5 and the use of tetraethyl lead as a fuel additive. One of the most toxic organometallic compounds ever discovered is $Ni(CO)_4$, which exists as a colourless volatile liquid. The threshold limit value (TLV), which is defined as the reasonable amount of the substance to which a person may be exposed to without adverse effects, of $Ni(CO)_4$ is 0.001 ppm. In comparison, hydrogen cyanide has a TLV of 4.7 ppm. Poisoning due to $Ni(CO)_4$ is caused by the destruction of lung tissue, and test animals that survive exposure develop cancer.

It has been suggested that nickel compounds may be a cause of cancer. Nickel is found in significant quantities in tobacco smoke, and thus may react with carbon monoxide and other compounds present in smoke to give volatile nickel compounds. Nickel compounds, which can catalyze organic cross coupling reactions at room temperature, would thus be introduced into the lungs, causing destruction of tissue or trigger cancer.

Nickel metal, which is used as a catalyst for hydrogenation of oils, inevitably gets incorporated into food products such as margarine and frying oils and all food cooked in such oils.

Bioorganometallics and Therapy

Although the first successful application of an organometallic compound as a drug is the anti-syphilis drug Salvarsan in 1910, it was only with the discovery of the anti-tumor properties of cisplatin in the late 20th century that research into organometallic based drugs took off. Since then, numerous advances have been made, and some successful examples of organometallic based drugs include ferroquine for the treatment of malaria and the use of radioactive 99mTc compounds as radiopharmaceuticals.

Figure: Structures of (clockwise from top): Salvarsan, cisplatin and ferroquine.

Currently, one of the most studied areas of the use of organometallic compounds as drugs is in the treatment of cancer. One in eight women living in western countries will be affected by breast cancer, one of the most deadly of all cancers affecting females. A majority of breast cancer tumors contain estrogen receptor α (ERα), and their growth is stimulated by estrogens. Treatment of these cancers involves using antiestrogens, and the most commonly used is tamoxifen, which is converted in vivo into its active form, hydrotamoxifen.

Figure: Structure of tamoxifen and hydrotamoxifen.

However some hormone dependent tumors do not respond to tamoxifen, and some will become resistant to the drug. There are also a group of tumors that contain a different form of estrogen receptors (ERβ), and are classified as hormone independent. Thus there is a need to search for better drugs.

One of the ideas is to use tamoxifen as a vector for organometallic fragments which may be potentially cytotoxic. Some examples include substitution of the β phenyl ring of tamoxifen with Cp_2TiCl_2 or $CpRe(CO)_3$ moieties.

Figure: Organotitanium and organorhenium derivatives of tamoxifen.

While the titanium derivative had an estrogenic effect, the rhenium derivative had antiestrogenic activity similar to that of hydrotamoxifen. A class of organometallic derivatives of tamoxifen that has shown a lot of promise is the ferrocifens.

R = H : Ferrocifens
R = OH : Hydroferrocifens

Figure General structure of the ferrocifens.

Not only do they have better antiproliferative effects on estrogen dependent cancer cells as compared with tamoxifen but they also show excellent cytotoxic effects against estrogen independent cancer cells, a feature that is absent in tamoxifen.16 While the exact mode of action of the ferrocifens is only known to an extent, the fact that it is about to enter phase 1 clinical trials demonstrates the potential of such bioorganometallic compounds as pharmaceuticals.

Bioorganometallics and Bioanalysis

The first effort into using an organometallic compound for bioanalysis was with a synthetic polypeptide which had ferrocene units that were labeled with radioactive ^{59}Fe. It was soon realized that the use of radioactive probes had some limitations, including limited half-life, relatively small amount of isotopes which could be used, cost, and serious health implications. This led to the development of other types of probes, which include fluorescent and metal probes for metalloimmunoassay. An example of an organometallic compound which has been used in a metalloimmunoassay is a manganese-nortryptiline tracer.

Figure: Structure of the cymantrenyl tracer of nortryptiline

A relatively newer technique involves using organometallic compounds as infrared (IR) probes. The carbonyl (CO) ligand, a commonly encountered ligand in organometallic compounds, exhibits strong stretching peaks in the 2000 – 1800 cm-1 region of the infrared spectrum. This region is free from interference by most organic functional groups, thus making CO containing organometallic complexes ideal as bioprobes. This was first demonstrated with lamb uterine cytosol which was incubated with an estradiol analogue derivatized with a $Cr(CO)_3$ moiety .

Figure: Structure of estradiol analogue derivatized with $Cr(CO)_3$

More recently, the use of a triosmium cluster as an IR tag for the imaging of oral mucosa cells has been successfully performed.23 The advantages of IR imaging include narrower line-widths and lower energies involved, while the advantages of using a triosmium cluster includes both air and water stability.

Biomineralization

The formation of inorganic materials with complex form is a widespread biological phenomenon (biomineralization) that occurs in almost all groups of organisms from prokaryotes (e.g., magnetite nanocrystals in certain bacteria) to humans (bone and teeth). Among the most spectacular examples of biomineralization are the intricately structured cell walls of diatoms, a large group of single-celled eukaryotic algae that are present in almost all water habitats. Diatom cell walls are made of amorphous, hydrated SiO_2 (silica) and exhibit highly regular porous patterns which are hierarchically arranged from the nano- to micrometer scale. The silica structures are produced by polycondensation of $Si(OH)_4$ (silicic acid) molecules which occurs in a specific intracellular compartment, termed the silica deposition vesicle (SDV). These complex biomineral structures reveal our limited understanding of a fundamental biological question: How does a cell translate linear DNA sequence information into patterned three-dimensional structures? Therefore, important lessons will be learned from diatoms regarding the mechanism by which eukaryotic cells assemble their cellular machinery to execute a genetically encoded morphogenetic program. Furthermore, owing to the structural intricacies and exceptional mechanical properties of biominerals, and their formation at mild (i.e., physiological) conditions, biomineralization is regarded as a paradigm for the development of novel routes for the synthesis of functional materials with nanometer precision in three dimensions (Bio-Nanotechnology).

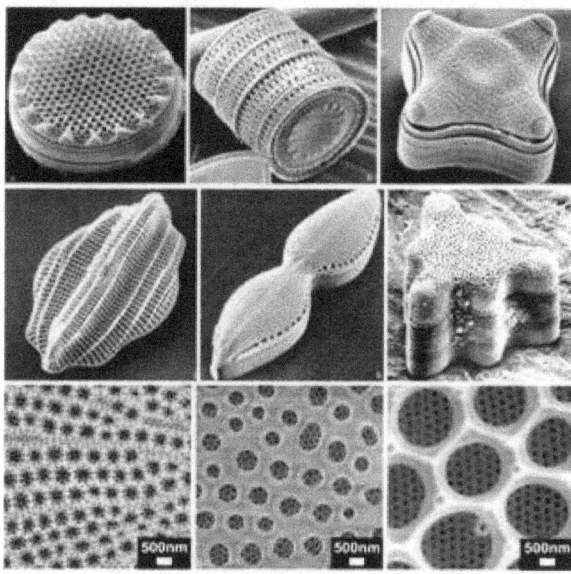

Scanning electron microscopy images of cell walls from different diatom species. Images in top and middle row show overviews and the bottom row shows details of individual cell walls.

The aim of the Kröger group is to understand the mechanism of silica biomineralization in diatoms by identifying and characterizing the biomolecules involved in this process, investigating their self-assembly properties, and analyzing their silica formation properties in vitro. This biochemical approach builds on previous research that has led to the identification of the first biomolecules involved in diatom silica formation, termed silaffins, silacidins and long-chain polyamines. In collaboration with the groups of Ginger Armbrust (University of Washington, Seattle) and Thomas Mock (University of East Anglia, U.K.) the Kröger group is involved in diatom genome projects, and utilizes bioinformatics approaches to identify new biomineralization proteins in diatom genomes.

To investigate the location and possible function of putative biomineralization proteins in vivo, the Kröger group has developed a genetic transformation system for *Thalassiosira pseudonana*, the model diatom for silica biomineralization studies. This allows for expression of GFP fusion proteins in vivo where we can follow their location in the cell during different stages of the cell cycle and silica morphogenesis.

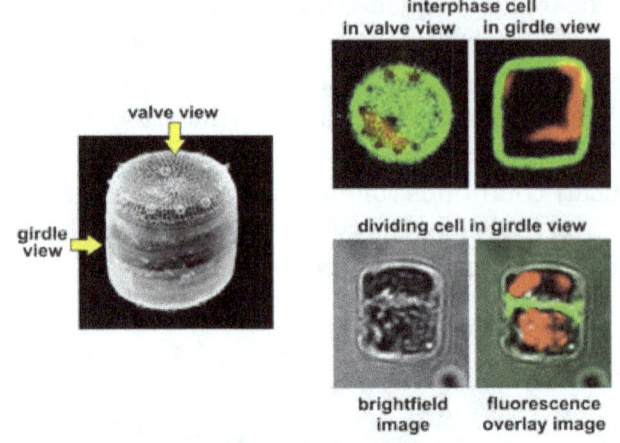

Expression of a GFP-tagged silaffin tpSil3 in *T. pseudonana.*

The top left image shows a scanning electron micrograph of *T. pseudonana.* The micrographs on the right show light microscopy images (brightfield, and confocal fluorescence microscopy) of *T. pseudonana* cells expressing a silaffin-GFP fusion protein. The red fluorescence is caused by the chloroplasts. In interphase cells the silaffin fusion protein is located in all parts of the cell wall (top). During cell division the silaffin-GFP fusion protein becomes incorporated into the newly forming valve part of the silica (bottom).

Iron Biomineralization

Many structures formed by living organisms are minerals. Examples include apatite [$Ca_2(OH)PO_4$] in bone and teeth, calcite or aragonite ($CaCO_3$) in the shells of marine organisms and in the otoconia (gravity device) of the mammalian ear, silica (SiO_2) in grasses and in the shells of small invertebrates such as radiolara, and iron oxides, such as magnetite (Fe_3O_4) in birds and bacteria (navigational devices) and ferrihydrite FeO(OH) in ferritin of mammals, plants, and bacteria. Biomineralization is the formation of such minerals by the influence of organic macromolecules, e.g., proteins, carbohydrates, and lipids, on the precipitation of amorphous phases, on the initiation of nucleation, on the growth of crystalline phases, and on the volume of the inorganic material.

Iron oxides, as one of the best-studied classes of biominerals containing transition metals, provide good examples for discussion. One of the most remarkable recent characterizations of such processes is the continual deposition of single-crystal ferric oxide in the teeth of chiton. Teeth of chiton form on what is essentially a continually moving belt, in which new teeth are being grown and moved forward to replace mature teeth that have been abraded. However, the study of the mechanisms of biomineralization in general is relatively recent; a great deal of the information currently available, whether about iron in ferritin or about calcium in bone, is somewhat descriptive.

Three different forms of biological iron oxides appear to have distinct relationships to the proteins, lipids, or carbohydrates associated with their formation and with the degree of crystallinity. Magnetite, on the one hand, often forms almost perfect crystals inside lipid vesicles of magneto-bacteria. Ferrihydrite, on the other hand, exists as large single crystals, or collections of small crystals, inside the protein coat of ferritin; however, iron oxides in some ferritins that have large amounts of phosphate are very disordered. Finally, goethite [α-FeO(OH)] and lepidocrocite [γ-FeO(OH)] form as small single crystals in a complex matrix of carbohydrate and protein in the teeth of some shellfish (limpets and chitons); magnetite is also found in the lepidocrocite-containing teeth. The differences in the iron-oxide structures reflect differences in some or all of the following conditions during formation of the mineral: nature of co-precipitating ions, organic substrates or organic boundaries, surface defects, inhibitors, pH, and temperature. Magnetite can form in both lipid and protein/carbohydrate environments, and can sometimes be derived from amorphous or semicrystalline ferrihydrite-like material (ferritin).

Synthetic iron complexes have provided models for two stages of ferritin iron storage and biomineralization.

1. The early stages, when small numbers of clustered iron atoms are bound to the ferritin protein coat, and

2. The final stages, where the bulk iron is a mineral with relatively few contacts to the protein coat. In addition, models have begun to be examined for the micro-environment inside the protein coat.

Among the models for the early or nucleation stage of iron-core formation are the binuclear Fe(III) complexes with $[Fe_2O(O_2CR_2)]^{2+}$ cores; the three other Fe(III) ligands are N. The μ-oxo complexes, which are particularly accurate models for the binuclear iron centers in hemerythrin, purple acid phosphatases, and, possibly, ribonucleotide reductases, may also serve as models for ferritin, since an apparently transient Fe(II)-O-Fe(III) complex was detected during the reconstitution of ferritin from protein coats and Fe(II). The facile exchange of (O_2CR) for (O_2PR) in the binuclear complex is particularly significant as a model for ferritin, because the structure of ferritin cores varies with the phosphate content. An asymmetric trinuclear $(Fe_3O)^{7+}$ complex and an $(FeO)_{11}$ complex have been prepared; these appear to serve as models for later stages of core nucleation (or growth).

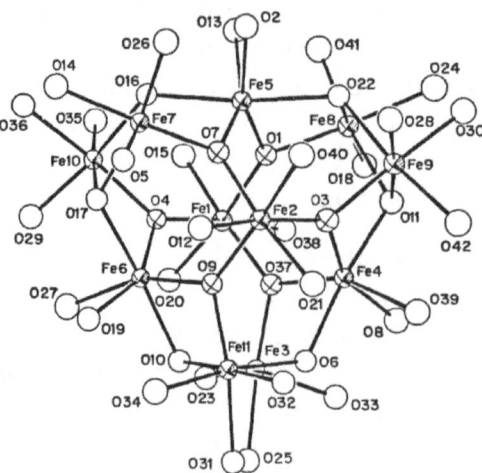

Figure: The structure of a model for a possible intermediate in the formation of the ferritin iron core. The complex consists of 11 Fe(III) atoms with internal oxo-bridges and a coat of benzoate ligands; the Fe atoms define a twisted, pentacapped trigonal prism.

Models for the full iron core of ferritin include ferrihydrite, which matches the ordered regions of ferritin cores that have little phosphate; however, the site vacancies in the lattice structure of ferrihydrite [FeO(OH)] appear to be more regular than in crystalline regions of ferritin cores. A polynuclear complex of iron and microbial dextran (α-1,4-D-glucose)$_n$ has spectroscopic (Mössbauer, EXAFS) properties very similar to those of mammalian ferritin, presumably because the organic ligands are similar to those of the protein (-OH, -COOH). In contrast, a polynuclear complex of iron and mammalian chondroitin sulfate (α-1,4-[α-1,3-D-glucuronic acid-N-acetyl-D-galactosamine-4-sulfate]$_n$) contains two types of domains: one like mammalian ferritin [FeO(OH)] and one like hematite (α-Fe$_2$O$_3$), which was apparently nucleated by the sulfate, emphasizing the importance of anions in the structure of iron cores. Finally, a model for iron cores high in phosphate, such as those from bacteria, is Fe-ATP (4:1), in which the phosphate is distributed throughout the polynuclear iron complex, providing an average of 1 or 2 of the 6 oxygen ligands for iron.

The microenvironment inside the protein coat of ferritin has recently been modeled by encapsulating ferrous ion inside phosphatidylcholine vesicles and studying the oxidation of iron as the pH is raised. The efficacy of such a model is indicated by the observation of relatively stable mixtures of Fe(II)/Fe(III) inside the vesicles, as have also been observed in ferritin reconstituted experimentally from protein coats and ferrous ion.

Models for iron in ferritin must address both the features of traditional metalprotein interactions and the bulk properties of materials. Although such modeling may be more difficult than other types of bioinorganic modeling, the difficulties are balanced by the availability of vast amounts of information on Fe-protein interactions, corrosion, and mineralization. Furthermore, powerful tools such as x-ray absorption, Mössbauer and solid state NMR spectroscopy, scanning electron and proton microscopy, and transmission electron microscopy reduce the number of problems encountered in modeling the ferritin ion core.

Construction of models for biomineralization is clearly an extension of modeling for the bulk phase of iron in ferritin, since the major differences between the iron core of ferritin and that of other iron-biominerals are the size of the final structure, the generally higher degree of crystallinity, and, at this time, the more poorly defined organic phases. A model for magnetite formation has been provided by studying the coulometric reduction of half the Fe^{3+} atoms in the iron core of ferritin itself. Although the conditions for producing magnetite have .yet to be discovered, the unexpected observation of retention of the Fe^{2+} by the protein coat has provided lessons for understanding the iron core of ferritin. Phosphatidyl choline vesicles encapsulating Fe^{2+} appear to serve as models for both ferritin and magnetite; only further investigation will allow us to understand the unique features that convert Fe^{2+} to [FeO(OH)], on the one hand, and Fe_3O_4, on the other.

Theories of Inorganic Chemistry

Some of the significant theories central to the development of inorganic chemistry include the crystal field theory, ligand field theory and the theory of molecular symmetry. This chapter discusses in extensive detail about these theories of inorganic chemistry.

Crystal Field Theory

Crystal field theory (CFT) describes the breaking of orbital degeneracy in transition metal complexes due to the presence of ligands. CFT qualitatively describes the strength of the metal-ligand bonds. Based on the strength of the metal-ligand bonds, the energy of the system is altered. This may lead to a change in magnetic properties as well as color. This theory was developed by Hans Bethe and John Hasbrouck van Vleck.

In Crystal Field Theory, it is assumed that the ions are simple point charges (a simplification). When applied to alkali metal ions containing a symmetric sphere of charge, calculations of bond energies are generally quite successful. The approach taken uses classical potential energy equations that take into account the attractive and repulsive interactions between charged particles (that is, Coulomb's Law interactions).

$$E \propto \frac{q_1 q_2}{r}$$

with

- E the bond energy between the charges and
- q_1 and q_2 are the charges of the interacting ions and
- r is the distance separating them.

This approach leads to the correct prediction that large cations of low charge, such as K^+ and Na^+, should form few coordination compounds. For transition metal cations that contain varying numbers of d electrons in orbitals that are NOT spherically symmetric, however, the situation is quite different. The shape and occupation of these d-orbitals then becomes important in an accurate description of the bond energy and properties of the transition metal compound.

When examining a single transition metal ion, the five d-orbitals have the same energy. When ligands approach the metal ion, some experience more opposition from the d-orbital electrons than others based on the geometric structure of the molecule. Since ligands approach from different directions, not all d-orbitals interact directly. These interactions, however, create a splitting due to the electrostatic environment.

For example, consider a molecule with octahedral geometry. Ligands approach the metal ion along the x, y, and z axes. Therefore, the electrons in the d_{z^2} and $d_{x^2-y^2}$ orbitals (which lie along these axes) experience greater repulsion. It requires more energy to have an electron in these orbitals than it would to put an electron in one of the other orbitals. This causes a splitting in the energy levels of the d-orbitals. This is known as crystal field splitting. For octahedral complexes, crystal field splitting is denoted by Δ_o (or Δ_{oct}). The energies of the dz2 and dx2−y2 orbitals increase due to greater interactions with the ligands. The d_{xy}, d_{xz}, and d_{yz} orbitals decrease with respect to this normal energy level and become more stable.

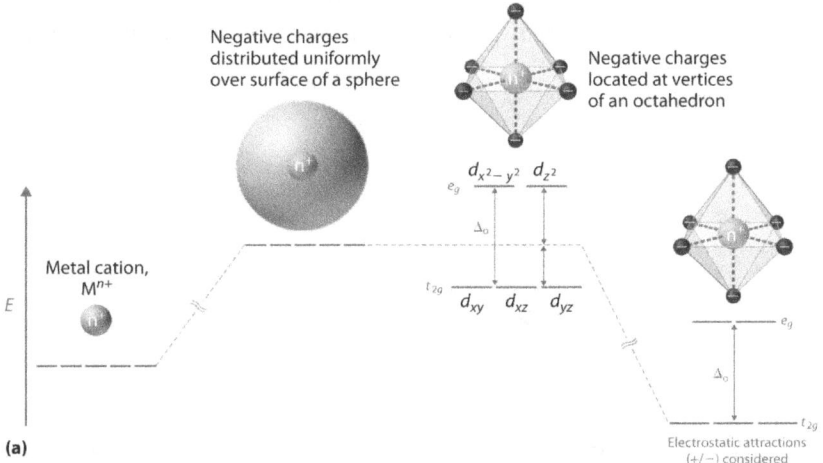

(a) Distributing a charge of −6 uniformly over a spherical surface surrounding a metal ion causes the energy of all five d orbitals to increase due to electrostatic repulsions, but the five d orbitals remain degenerate. Placing a charge of −1 at each vertex of an octahedron causes the d orbitals to split into two groups with different energies: the $d_{x^2-y^2}$ and d_{z^2} orbitals increase in energy, while the, d_{xy}, d_{xz}, and d_{yz} orbitals decrease in energy. The average energy of the five d orbitals is the same as for a spherical distribution of a −6 charge, however. Attractive electrostatic interactions between the negatively charged ligands and the positively charged metal ion (far right) cause all five d orbitals to decrease in energy but does not affect the splittings of the orbitals. (b) The two e$_g$ orbitals (left) point directly at the six negatively charged ligands, which increases their energy compared with a spherical distribution of negative charge. In contrast, the three t_{2g} orbitals (right) point between the negatively charged ligands, which decreases their energy compared with a spherical distribution of charge.

Electrons in Orbitals

According to the Aufbau principle, electrons are filled from lower to higher energy orbitals. For the octahedral case above, this corresponds to the d_{xy}, d_{xz}, and d_{yz} orbitals. Following Hund's rule, electrons are filled in order to have the highest number of unpaired electrons. For example, if one had a d^3 complex, there would be three unpaired electrons. If one were to add an electron, however, it has the ability to fill a higher energy orbital (d_{z^2} or $d_{x^2-y^2}$) or pair with an electron residing in the d_{xy}, d_{xz}, or d_{yz} orbitals. This pairing of the electrons requires energy (spin pairing energy). If the pairing energy is less than the crystal field splitting energy, Δ_o, then the next electron will go into the d_{xy}, d_{xz}, or d_{yz} orbitals due to stability. This situation allows for the least amount of unpaired electrons, and is known as low spin. If the pairing energy is greater than Δ_0, then the next electron will go into the d_{z^2} or $d_{x^2-y^2}$ orbitals as an unpaired electron. This situation allows for the most number of unpaired electrons, and is known as high spin. Ligands that cause a transition metal to have a small crystal field splitting, which leads to high spin, are called weak-field ligands. Ligands that produce a large crystal field splitting, which leads to low spin, are called *strong field ligands*.

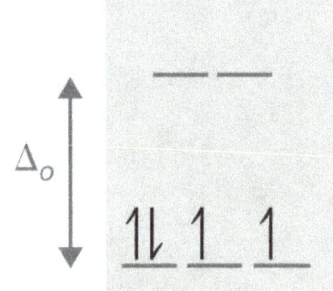

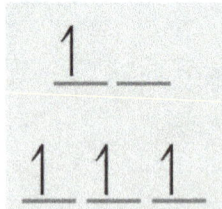

Low Spin, Strong Field ($\Delta_o > P$)   High Spin, Weak Field ($\Delta_o < P$)

Figure: Splitting for a d4d4 complex under a strong field (left) and a weak field (right). The strong field is a low spin complex, while the weak field is a high spin complex.

As mentioned above, CFT is based primarily on symmetry of ligands around a central metal/ion and how this anisotropic (properties depending on direction) ligand field affects the metal's atomic orbitals; the energies of which may increase, decrease or not be affected at all. Once the ligands' electrons interact with the electrons of the *d*-orbitals, the electrostatic interactions cause the energy levels of the d-orbital to fluctuate depending on the orientation and the nature of the ligands. For example, the oxidation state and the strength of the ligands determine splitting; the higher the oxidation state or the stronger the ligand, the larger the splitting. Ligands are classified as strong or weak based on the spectrochemical series:

$$I^- < Br^- < Cl^- < \underline{S}CN^- < F^- < OH^- < ox^{2-} < \underline{O}NO^- < H_2O < SCN^- < EDTA^{4-} < NH_3 < en < NO_2^- < CN^-$$

Note that SCN⁻ and NO_2- ligands are represented twice in the above spectrochemical series since there is two different Lewis base sites (e.g., free electron pairs to share) on each ligand (e.g., for SCN⁻ the electron pair on the sulfur or the oxygen can form the coordinate covalent bond to a metal). The specific atom that binds in such ligands is underlined.

In addition to octahedral complexes, two common geometries observed are that of tetrahedral and square planar. These complexes differ from the octahedral complexes in that the orbital levels are raised in energy due to the interference with electrons from ligands. For the tetrahedral complex, the d_{xy}, d_{xz}, and d_{yz} orbitals are raised in energy while the d_{z^2}, $d_{x^2-y^2}$ orbitals are lowered. For the square planar complexes, there is greatest interaction with the $d_{x^2-y^2}$ orbital and therefore it has higher energy. The next orbital with the greatest interaction is d_{xy}, followed below by d_{z^2}. The orbitals with the lowest energy are the d_{xz} and d_{yz} orbitals. There is a large energy separation between the d_{z^2} orbital and the d_{xz} and d_{yz} orbitals, meaning that the crystal field splitting energy is large. We find that the square planar complexes have the greatest crystal field splitting energy compared to all the other complexes. This means that most square planar complexes are low spin, strong field ligands.

Description of *d*-Orbitals

To understand CFT, one must understand the description of the lobes:

- d_{xy} : lobes lie in-between the x and the y axes.

- d_{xz} : lobes lie in-between the x and the z axes.

- d_{yz} : lobes lie in-between the y and the z axes.

- $d_{x^2-y^2}$: lobes lie on the x and y axes.

- d_{z^2} : there are two lobes on the z axes and there is a donut shape ring that lies on the xy plane around the other two lobes.

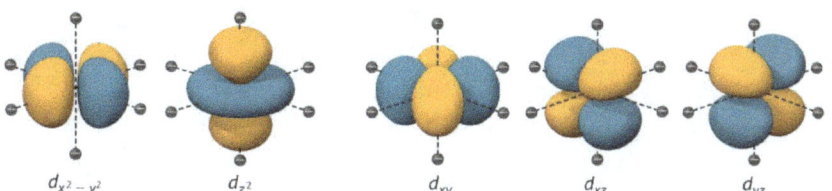

Figure: *Spatial arrangement of ligands in* the an octahedral ligand field with respect to the five d-orbitals.

Octahedral Complexes

In an octahedral complex, there are six ligands attached to the central transition metal. The d-orbital splits into two different levels. The bottom three energy levels are named d_{xy}, d_{xz}, and d_{yz} (collectively referred to as t_{2g}). The two upper energy levels are named $d_{x^2-y^2}$, and d_{z^2} (collectively referred to as e_g).

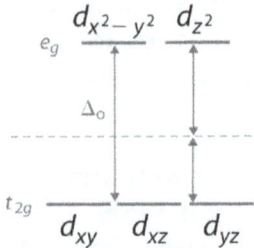

Figure: Splitting of the degenerate d-orbitals (without a ligand field)
due to an octahedral ligand field shown in Figure.

The reason they split is because of the electrostatic interactions between the electrons of the ligand and the lobes of the d-orbital. In an octahedral, the electrons are attracted to the axes. Any orbital that has a lobe on the axes moves to a higher energy level. This means that in an octahedral, the energy levels of e_g are higher $(.6\Delta_o)$ while t_{2g} is lower $(0.4\Delta_o)$. The distance that the electrons have to move from t_{2g} from e_g and it dictates the energy that the complex will absorb from white light, which will determine the color. Whether the complex is paramagnetic or diamagnetic will be determined by the spin state. If there are unpaired electrons, the complex is paramagnetic; if all electrons are paired, the complex is diamagnetic.

Tetrahedral Complexes

In a tetrahedral complex, there are four ligands attached to the central metal. The d orbitals also split into two different energy levels. The top three consist of the d_{xy}, d_{xz}, and d_{yz} orbitals. The bottom two consist of the $d_{x^2-y^2}$ and d_{z^2} orbitals. The reason for this is due to poor orbital overlap between the metal and the ligand orbitals. The orbitals are directed on the axes, while the ligands are not.

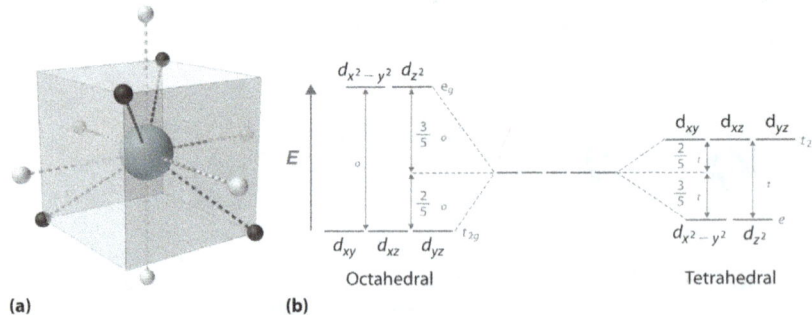

(a)   (b)

Figure: (a) Tetraheral ligand field surrounding a central transition metal (blue sphere).
(b) Splitting of the degenerate d-orbitals (without a ligand field) due to an octahedral ligand field
(left diagram) and the tetrahedral field (right diagram).

The difference in the splitting energy is tetrahedral splitting constant (Δ_t), which less than (Δ_o) for the same ligands:

$$\Delta_t = 0.44\Delta_o$$

Consequentially, Δ_t is typically smaller than the spin pairing energy, so tetrahedral complexes are usually high spin.

Square Planar Complexes

In a square planar, there are four ligands as well. However, the difference is that the electrons of the ligands are only attracted to the xy plane. Any orbital in the xy plane has a higher energy level. There are four different energy levels for the square planar (from the highest energy level to the lowest energy level): $d_{x^2-y^2}$, d_{xy}, d_{z^2}, and both d_{xz} and d_{yz}.

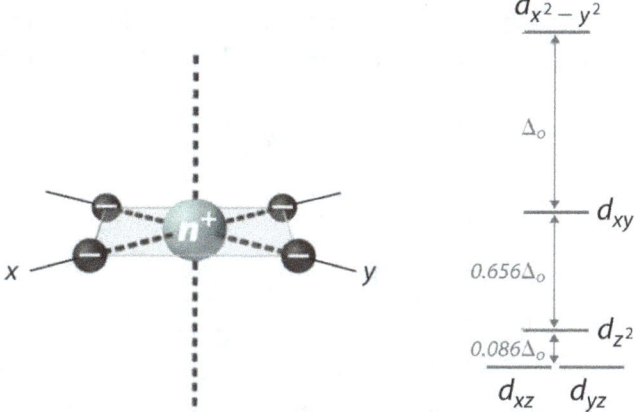

Figure: Splitting of the degenerate d-orbitals (without a ligand field) due to an square planar ligand field.

The splitting energy (from highest orbital to lowest orbital) is Δ_{sp} and tends to be larger then Δ_o.

$$\Delta_{sp} = 1.74\Delta_o$$

Moreover, Δ_{sp} is also larger than the pairing energy, so the square planar complexes are usually low spin complexes.

Example

For the complex ion $\left[Fe(Cl)_6\right]^{3-}$ determine the number of d electrons for Fe, sketch the d-orbital energy levels and the distribution of d electrons among them, list the number of lone electrons, and label whether the complex is paramagnetic or diamagnetic.

Solution

- Step 1: Determine the oxidation state of Fe. Here it is Fe^{3+}. Based on its electron configuration, Fe^{3+} has 5 d-electrons.

- Step 2: Determine the geometry of the ion. Here it is an octahedral which means the energy splitting should look like:

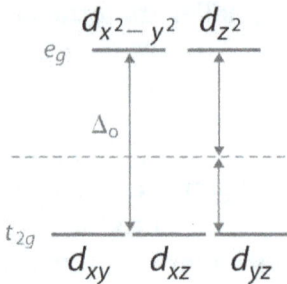

- Step 3: Determine whether the ligand induces is a strong or weak field spin by looking at the spectrochemical series. Cl⁻ is a weak field ligand (i.e., it induces high spin complexes). Therefore, electrons fill all orbitals before being paired.

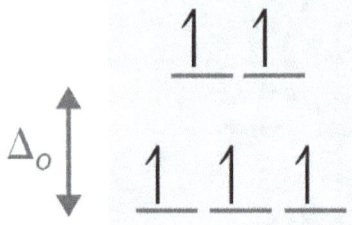

- Step four: Count the number of lone electrons. Here, there are 5 electrons.

- Step five: The five unpaired electrons means this complex ion is paramagnetic (and strongly so).

Example

A *tetrahedral* complex absorbs at 545 nm. What is the respective octahedral crystal field splitting ($\Delta o \Delta o$)? What is the color of the complex?

Solution

$$\Delta_t = \frac{hc}{\lambda}$$

$$\Delta_t = \frac{hc}{\lambda} \; \Delta_t = \frac{(6.626 \times 10^{-34} J \cdot s)(3 \times 10^8 m/s)}{545 \times 10^{-9} m} = 3.65 \times 10^{-19} J$$

However, the tetrahedral splitting (Δ_t) is $\sim 4/9$ that of the octahedral splitting (Δ_o).

$$\Delta_t = 0.44 \Delta_o$$

$$\Delta_o = \frac{\Delta_t}{0.44} = \frac{3.65 \times 10^{-19} J}{0.44} = 8.30 \times 10^{-18} J$$

This is the energy needed to promote one electron in one complex. Often the crystal field splitting is given per mole, which requires this number to be multiplied by Avogadro's Number (6.022×10^{23}).

This complex appears red, since it absorbs in the complementary green color (determined via the color wheel).

Ligand Field Theory

Concepts from molecular orbital theory are useful in understanding the reactivity of coordination compounds. One of the basic ways of applying MO concepts to coordination chemistry is in Ligand Field Theory. Ligand field theory looks at the effect of donor atoms on the energy of d orbitals in the metal complex.

There are two ways in which we sometimes think about the effect of ligands on the d electrons on a metal. On the basis of simple electron-electron repulsion, donation of a lone pair might raise an occupied d orbital in energy. Alternatively, we can think about bonding interactions between ligand orbitals and d orbitals. This second way of thinking about things is a little bit more useful, and that's the approach we'll focus on, here.

Either way, there are interactions between ligand electrons and d electrons, that usually end up raising the d electrons in energy. The effect depends on the coordination geometry geometry of the ligands. Ligands in a tetrahedral coordination sphere will have a different effect than ligands in an octahedral coordination sphere, because they will interact with the different d orbitals in different ways.

- Ligand field theory looks at the effect of donor atoms on the energy of d orbitals in the metal complex.

- The effect depends on the coordination geometry geometry of the ligands.

Octahedral Case

Suppose a complex has an octahedral coordination sphere. Assume the six ligands all lie along the x, y and z axes.

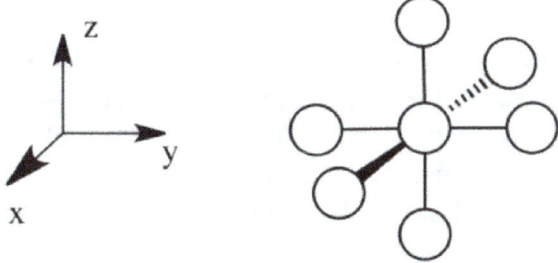

There are two d orbitals that will interact very strongly with these ligands: the d^2_x, y_2, which lies directly on the x and y axes, and the d^2_z , which lies directly on the z axis. Together, these two metal orbitals and the ligand orbitals that interact with them will form new bonding and antibonding orbitals.

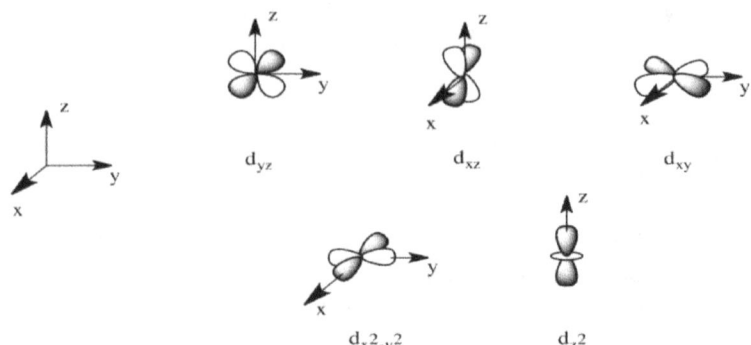

The drawing below is simplified. The ligands will also interact with s and p orbitals, but for the moment we're not going to worry about them. We also won't worry about interactions from the other four ligands with the d orbitals (possible by symmetry considerations, but also a more complicated picture).

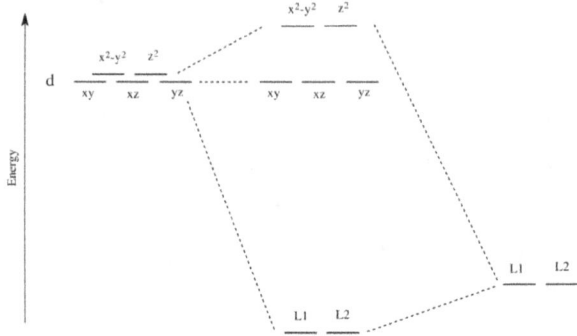

Now, remember that metals usually have d electrons that are much higher in energy than those on typical donor atoms (like oxygen, sulfur, nitrogen or phosphorus). That means the antibonding combinations will be much closer in energy to the original d orbitals, because both are relatively high in energy. The bonding combination will be much closer in energy to the original ligand orbitals, because these ones are all relatively low in energy.

That energetic similarity generally translates into a similarity in shape and location as well. In other words, the antibonding combination between a d orbital and a ligand orbital is a lot like the original d orbital. The bonding combination is more like the original ligand orbital than the original d orbital. Because of those similarities, inorganic chemists often refer to those antibonding orbitals as if they were still the original d orbitals.

These two orbitals will be raised relatively high in energy by sigma bonding interactions with the donor orbitals. If there are electrons in the picture, it might look something like this:

- Assume the six ligands all lie along the x, y and z axes.

- The d_{x^2,y_2} and the d_z^2 orbitals lie along the bond axes.

- These two orbitals will be raised relatively high in energy.

- These orbitals are like antibonding levels.

- These orbitals are sometimes called the "eg" set of orbitals. The term "eg" comes from the mathematics of symmetry.

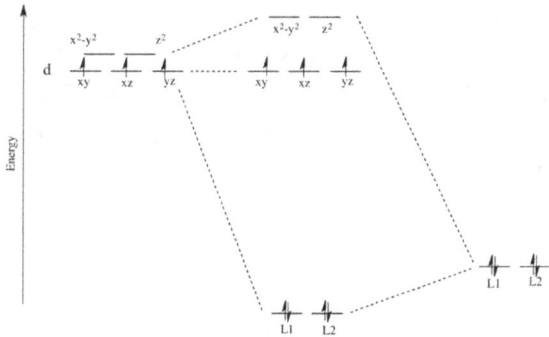

On the other hand, the other three d orbitals, the d_{xy}, d_{xz} and d_{yz}, all lie between the donor ligands, rather than hitting them head-on. These orbitals will interact less strongly with the donor electrons.

- The dxy, dxz and dyz orbitals all lie between the bond axes.

- These three orbitals will be changed in energy only a little.

- These orbitals are more like non-bonding orbitals.

- These orbitals are sometimes called the "t_{2g}" set of orbitals.

Remember, only the energy of the electrons affects the overall energy of the system. The unoccupied d orbitals are raised in energy, but the occupied orbitals go down in energy (or else stay the same).

Apart from the stabilization of the complex, there is another consequence of this picture. What we are left with is two distinct sets of d energy levels, one lower than the other. That will have an effect on the electron configuration at the metal atom in the complex. That means there will be cases where electrons could be paired or unpaired, depending on how these orbitals are occupied.

Take the case of the biologically important iron(II) ion. It has a d6 valence electron configuration. In less formal parlance of inorganic chemistry, "iron(II) is d^6 ". In an iron(II) ion all alone in space, all the d robitals would have the same energy level. We would put one electron in each orbital, and have one left. It would need to pair up in one of the d orbitals. (Notice that, in the chemistry of transition metalions, the valence s and p orbitals are always assumed to be unoccupied).

Things are very different in an octahedral complex, like $K_4[Fe(CN)_6]$. In that case, the d orbitals are no longer at the same energy level. There are two possible configurations to consider.

In one case, one electron would go into each of the lower energy d orbitals. A choice would be made for the fourth electron. Does it go into the higher energy d orbital, or does it pair up with one of the lower energy d electrons? The choice depends on how much higher in energy the upper d orbitals are, compared to how much energy it costs to put two electrons in the same d orbital.

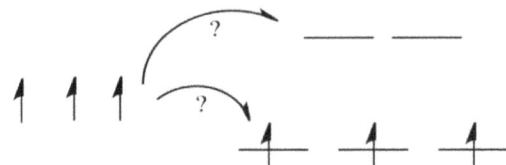

If the "d orbital splitting energy" is pretty low, so that the two sets of d orbitals are still pretty similar in energy, the next electron can go into a higher orbital. Pairing would not be required until the final electron. Overall, that would leave four unpaired electrons, just like in the case of a lone metal ion in space. This is called the "high-spin" case, because electrons can easily go into the higher orbital.

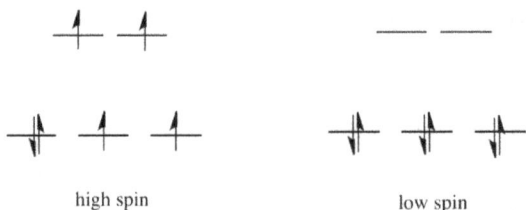

high spin   low spin

If the d orbital splitting energy is too high, the next electron must pair up in a lower orbital. All three remaining electrons pair up, and so there are no unpaired electrons in the complex. This is called the "low-spin" case, because electrons more easily pair up in the orbital.

So the overall rule is that if the energy to pair up the electrons is greater than the energy needed to get to the next level, the electron will go ahead and occupy the next level.

However, if the energy it takes to get to the next level is more than it would cost to pair up, the electrons will just pair up instead.

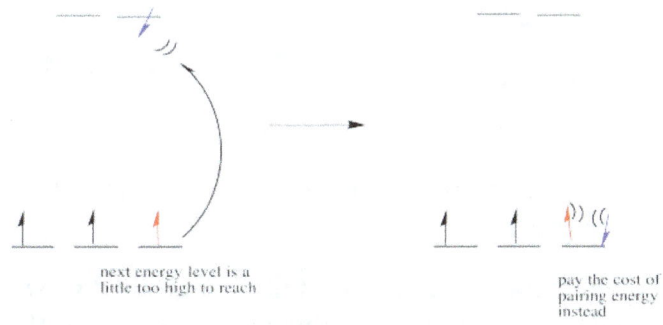

The electron configuration can be "high spin" or "low-spin", depending on how large the energy splitting is between the two sets of d orbitals.

The difference between the high-spin case and the low-spin case is significant, because unpaired electrons affect the magnetic properties of a material. The lowspin case would be diamagnetic, resulting in no interaction with a magnetic field. However, the high-spin case would be paramagnetic, and would be attracted to a magnetic field.

It turns out $K_4[Fe(CN)_6]$ is diamagnetic. Thus, it is pretty clear that it is a low-spin complex. The energy difference between the two d orbital levels is relatively large in this case.

In addition to influencing magnetic properties, whether a complex is high- or lowspin also influences reactivity. Compounds with high-energy d electrons are generally more labile, meaning they let go of ligands more easily.

- Electron configuration influences magnetic properties

- Electron configuration influences lability (how easily ligands are released)

Reasons for Low-spin vs. High-spin: The Effect of the Metal Ion

There are a few factors that determine the magnitude of the d orbital splitting, and whether an electron can occupy the higher energy set of orbitals, rather than pairing up. It is based partly on ligand field strength, which is explored on the next page. It also depends on the charge on the metal ion, and whether the metal is in the first, second or third row of the transition metals.

The higher the charge on the metal, the greater the splitting between the d orbital energy levels. For example, Fe(II) is usually high spin. It has a smaller splitting between the lower and higher d orbital levels, so electrons can more easily go to the higher level rather than pair up on the lower level.

On the other hand, Fe(III) is usually low spin. It has a larger splitting between the d levels. In that case, it costs less energy for electrons to pair up in the lower level than to go up to the higher level.

- High-spin versus low-spin cases involve a trade-off between the d orbital splitting energy and the pairing energy.

- 2nd and 3rd row transition metals are usually low spin.

- 1st row transition metals are often high spin.

- However, 1st row transition metals and be low spin if they are very positive (usually 3+ or greater).

There is a lot going on in metal ions, but we'll take a simplified view of things. Thinking only about electrostatics, we can try to imagine what happens to those electrons when the charge on the metal ion changes.

First we need to know about Coulomb's law. Coulomb's law states that the force of attraction between the electron and the nucleus depends on only two factors: the amount of positive charge in the nucleus, and the distance between the nucleus and the electron.

The greater the charge on the nucleus, the greater the attraction between the electron and the nucleus.

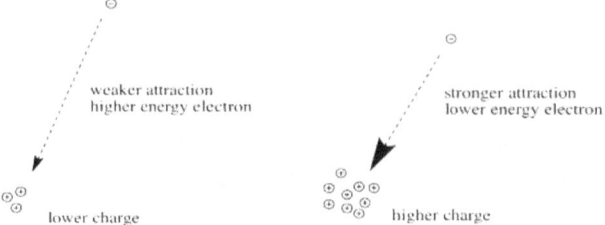

The farther an electron is from the nucleus, the weaker the attraction between the electron and the nucleus.

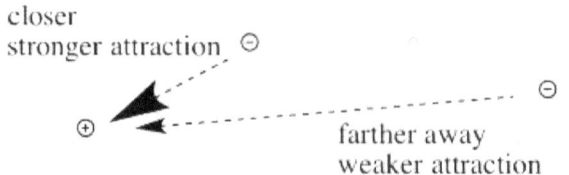

Coulomb's law can be used to evaluate the potential energy of the electron. It is one of the factors that determines how high or low those electronic energy levels are that we see in energy level diagrams for atoms, ions and molecules. The energy of the electron varies in a roughly similar way: the greater the charge on the nucleus, the lower the energy of the electron. Also, the closer the electron is to the nucleus, the lower its energy.

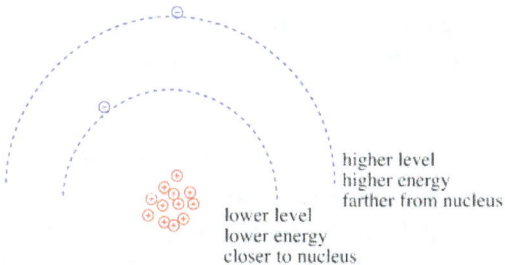

Roughly speaking, electrons at higher energy are farther from the nucleus. Electrons at lower energy are closer to the nucleus.

What happens if the charge increases? Maybe a lot more protons are added to the nucleus. Maybe some electrons are lost, so that to the remaining electrons it just feels like the charge of the nucleus has increased. Then the electrons should be more attracted to the nucleus. They get a little closer. Their potential energy drops.

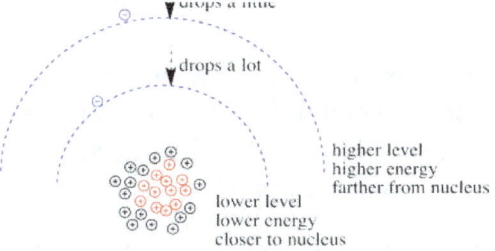

Of course, if one electron is closer to the nucleus already, it feels that increase in positive charge more strongly than an electron that is farther away. Consequently it drops further in energy than an electron that is further away.

If we translate that idea into a picture of the d orbital energy levels in an octahedral geometry, it looks like this:

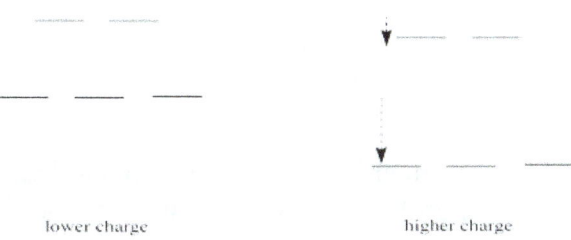

When the charge on the metal ion is increased, both the higher and the lower levels drop in energy. However, the lower level drops more. Thus, the gap between the levels gets wider.

Metals in the second and third row of the periodic table almost never form high-spin complexes. The d orbital energy splitting in these cases is larger than for first row metals. From a very simple point of view, these metals have many more protons in their nuclei than the first row transition metals, dropping that lower set of d electrons lower with respect to the higher set.

That isn't the whole picture for the second and third row transition metals, however. Remember, we are simplifying, and there are factors we won't go into. However, it is important to know that metal-ligand bond strengths are much greater in the second and third row than in the first. We'll look at the whole interaction diagram for an octahedral complex now, including contributions form metal s and p orbitals.

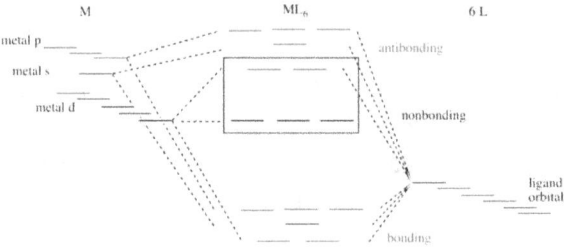

Like all ligand-metal interaction diagrams, the energy levels of the ligands by themselves are shown on one side. The metal's electronic energy levels are shown on the other side. The result of their interaction, a metal-ligand complex, is shown in the middle. The d orbital splitting diagram is shown in a box.

Suppose the diagram above is for a first row transition metal. The diagram for a second or third row metal is similar, but with stronger bonds.

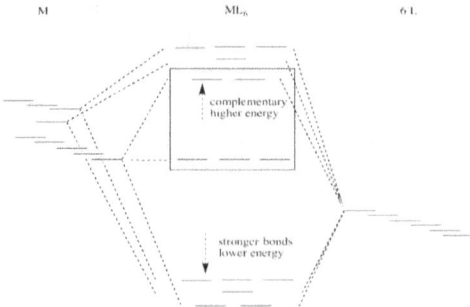

If the bonding interaction is stronger between the metal and ligand, then so is the antibonding interaction. The antibonding levels are bumped higher in energy as the bonding levels sink lower. Generally that's OK, because when the electrons are filled in, they will be found preferentially at the lower levels, not the higher ones. There will be a net lowering of electronic energy.

Why do second and third row transition metals form such strong bonds? Bond strengths are very complicated. In general, there is greater covalency between these metals and their ligands because of increased spatial and energetic overlap. Rather than go into those factors, we'll just think about all those extra protons in the nucleus that are attracting the ligand electrons more strongly.

There is one more important distinction that makes second and third row transition metals low spin. In addition, the pairing energy is lower in these metals because the orbitals are larger. There is more room for two electrons in one orbital, with less repulsion. As a result, electrons are much more likely to pair up than to occupy the next energy level.

- 2nd and 3rd row transition metals have stronger bonds, leading to a larger gap between d orbital levels

- 2nd and 3rd row transition metals have more diffuse orbitals, leading to a lower pairing energy

It is significant that most important transition metal ions in biology are from the first row of the transition block and are pretty labile. That fact plays an important role in the ease of formation and deconstruction of transition-metal containing proteins. In terms of formation, if the metal is more easily released by its previous ligands (either water or some compound that delivers the metal to the site of protein construction), it can form the necessary protein more quickly. However, even if a metal-containing enzyme plays a useful role, it should not be too stable, because we need to be able to regulate the level of protein concentration for optimum activity, or disassemble protein if it becomes damaged. Thus, it is important that the metal ion can be removed easily.

Theory of Molecular Symmetry

In everyday life we think of it as a spacial regularity. Mathematically, an object is symmetric, if it is does not change for certain transformations.

For example:

- A symmetric building will not change if we reflect it on the palne cutting it in the middle.

- Merry-go-round can be rotated such that one seat gets to an other, nothing changes.

- An egg can be rotated by any angle about its axis, we will not see any change.

Molecules also own similar properties. For example:

- Water molecule can be reflected on the plane cutting the oxygen atom and intersecting in the middle of the two hydrogen atoms. Also, we can rotate it by 180 degrees around the axis going through the oxygen atom.

- In case of ammonia, one can rotate it by 120 degrees around the axis going through the nitrogen and middle of the triangle defined by the three hydrogen atoms. Also, there are three planes for reflection including the oxygen and one of the hydrogens.

Let us gather all possible symmetry operations:

- C_n – rotation around an axis through an angel of $2\pi/n$ radian (gir)

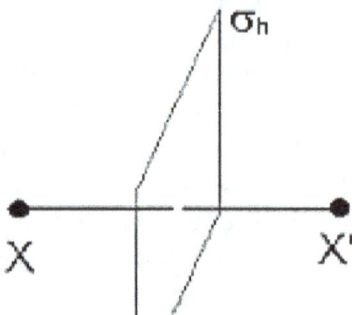

- σ – reflection on a mirror plane (special cases: σ_v, σ_h, σ_d)

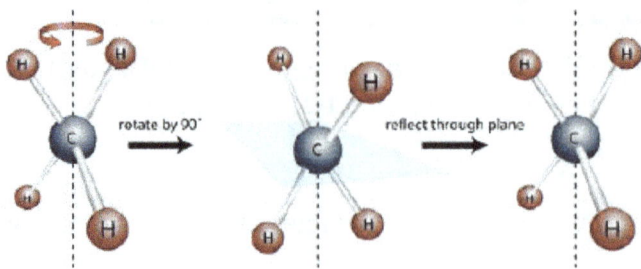

- S_n – improper rotation (rotation-reflection): a combination of C_n rotation és σ_h reflection (plane is orthogonal to the axis) as an individual operation (giroid)

- i – inversion, i.e. reflection on a point ($i = S_2$)

- E – identity operation; leaves the object unchanged, only for mathematical purposes

Point Groups

Now we can return to the question: what is symmetry? Symmetry is the collection of symmetry operations which do not change the object.

Chemical examples:

- Water: $C_2(z)$, σ_{zx}, σ_{zy}, E

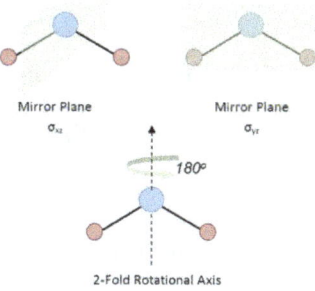

- Ammonia: $C_3(z)$, 3 times σ_v, E

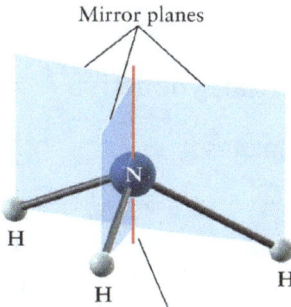

- Benzene: C_6, 6 times C_2, σh (horizontal, perpendicular to main axis), 6 times σ_v, i, and some more.

- Formaldehyde: $C_2(z)$, σ_{zx}, σ_{zy}, E

One can conclude that the operators not modifying the nuclear frame characterize the molecules. These are characteristic for the symmetry, since, for example, the same operators are there for water and formaldehyde. Such group of operators are called point group. (The name comes from the fact that these operators form a "group" in mathematical sense.) Point groups are labeled by the so called Schönflies-symbols:

- C_n: the group including the C_n operations (special case: C_1 no symmetry at all);

- C_{nv}: the group including the C_n, as well as σv mirror plane;

- C_{nh}: the group includes the C_n, as well as σh mirror plane;

- D_n: the group includes C_n as well as n times C_2 rotations which are perpendicular to the main axis;

- D_{nh}: same as D_n but also includes mirror planes perpendicular to the main axis;

- D_{nd}: same as D_n but also includes mirror planes including the main axis;

- S_n: the group includes S_n operation;

- T_d: group having tetrahedron symmetry;

- $C_{\infty v}$: this group includes a rotation through any angle ($n = \infty$);

- $D_{\infty h}$: this group includes a rotation through any angle; ($n = \infty$) and, in addition, mirror plan perpendicular to this axis;

- O_3^+: point group for spherical symmetry.

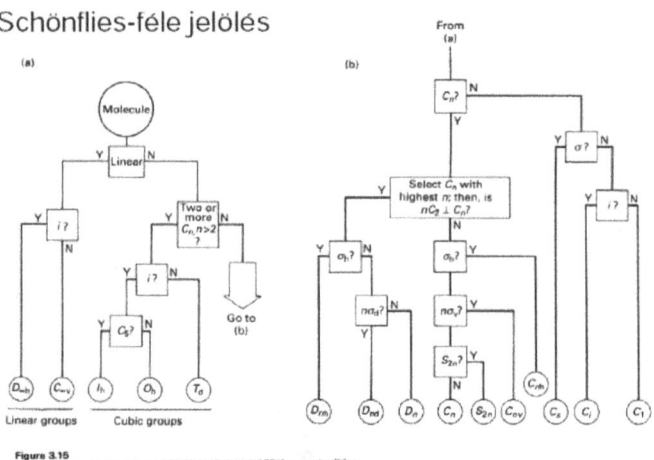

Schönflies-féle jelölés

Figure 3.15
Shriver, Atkins, and Langford: *INORGANIC CHEMISTRY*, second edition
©1990, 1994 D. F. Shriver, P. W. Atkins, and C. H. Langford
W. H. Freeman and Company

Chemical example molecule	symmetry operation	point group
water	C_2, σ_{zx}, σ_{zy}, E	C_{2v}
ammonia	$C_3(z)$, 3 times σ_v, E	C_{3v}
benzene	C_6, 6 *times* C_2, σ_h, 6 *times* σv, i, etc.	D_{6h}

fomaldehyde	$C_2(z),\ \sigma_{zx},\ \sigma_{zy},\ E$	C_{2v}
ethylene		D_{2h}
acetilene		$D_{\infty h}$
carbon monoxide		$C_{\infty v}$

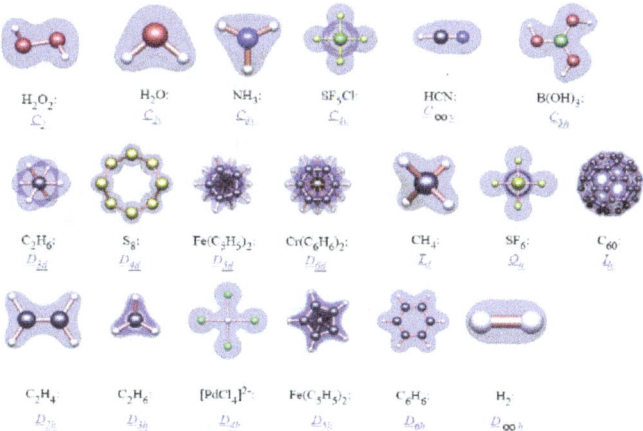

Representation, Character Table

We have seen that the operators of the molecule's point group do not change the molecular frame. It follows that the wave function of the molecule can not change either, or eventually it can change sign. Again, the explanation of the minus sign can be given by the physical meaning of the square of the wave function.

We can classify the wave function according to the sign caused by the symmetry operation. But how many possibilities are there? Look at again water as an example!

Sorszám	E	C_2	σ_{zx}	σ_{zy}
1	1	1	1	1
2	1	1	-1	-1
3	1	-1	1	-1
4	1	-1	-1	1

There are no other possibility, since certain combinations are excluded. Therefore we can state that the wave function of water can belong to only four different classes with symmetry properties according to the table above. These correspond to one of the rows of the above table. These „classes" represent the point group, therefore are termed „representation". In particular, these are the elementary representations meaning that they can not be reduced further, therfore these are called as irreducible representations.

The irreducible representations and their properties (signs resulting out of the symmetry operations performed on them) are collected in the so called character tables. The character table of the C_{2v} point group is given by:

C_{2v}	E	C_2	σ_{zx}	σ_{zy}
A_1	1	1	1	1
A_2	1	1	-1	-1
B_1	1	-1	1	-1
B_2	1	-1	-1	1

Notice that above we have constructed this table by putting in the possible signes. In case of „very symmetric" systems the irreps are not one-dimensional, certain symmetry operation can be described only by two or more functions. As example we mention ammonia (C_{3v} point group).

C_{3v}	E	$2C_3$	$3\sigma_v$
A_1	1	1	1
A_2	1	1	-1
E	2	-1	0

If a point group has many dimensional irreps, then the system will have degeneracy, the manyfold of the degeneracy is the same as the dimension of the corresponding irrep. In the above example:

- Ammonia will have double degenerate energy levels;

- Water does not have degeneracy;

- In atoms there are single-fold, triple-fold, quintuple-fold, degeneracies, since the O_3^+ : point group has irreps of one, three, five, etc. dimension.

Some Nomenclature

Character: the elements of the character table are called „character"; these correspond to the symmetry operation of their column and irrep of their row. In one dimensional case this is very expressive: if the character is 1, the irrep is symmetric, if -1, the irrep is antisymmetric with respect to the given operation.

Symboles denoting the irreps: these are given in the first row of the character table. If this letter is A or B, the irrep is one dimensional, in case E it is two-dimensional, in case of F and T three-dimensional. Exception: $C_{\infty v}$ and $D_{\infty h}$ groups, where the one-dimensional irrep is called Σ, the two-dimensional is Π and Δ, etc.

Total symmetric representation: in this case character for all operations is 1, i.e. such function is symmetric for all operations, and it does not change sign. Its symbole usually includes the letter A1, but in case of $D_{\infty h}$ point group its is denoted by Σ_g^+. Dimension of the representation: given by the character of the identity operator (E).

6

Techniques to Characterize Inorganic Compounds

The bulk properties of inorganic compounds, their optical and electronic properties are studied using diverse techniques. Some common techniques used are spectroscopy, X-ray crystallography, electrochemistry and dual polarization interferometry, among others. All such important techniques for the characterization of inorganic compounds have been covered in this chapter.

Spectroscopy

Spectroscopy is the study of the interaction of electromagnetic radiation with matter. Spectroscopy has many applications in the modern world, ranging from nondestructive examination of materials to medical diagnostic imaging (e.g., MRIs, CAT scans). In a chemical context, spectroscopy is used to study energy transitions in atoms and molecules. The transitions are interpreted and can serve to identify the molecule or give clues about the molecular structure. Spectroscopy is a powerful tool for inorganic chemists to help identify the compounds that have been prepared. Problem solving plays a crucial role in the interpretation of spectra, and you will find that your deductive reasoning skills will be challenged as you apply the principles of spectroscopy to solving chemical problems.

When a molecule interacts with electromagnetic radiation, energy is absorbed and the molecule is promoted, or is said to undergo a transition, to a higher energy state (excited state). In order for absorption to occur, the energy of the radiation must match the energy difference between the quantized energy levels of the molecule. For example, in figure, E_1 and E_2 are the quantized energy levels and ΔE is the energy difference ($\Delta E = E_2 - E_1$) that must match the energy of the incident radiation.

$$\Delta E = hv = \frac{hc}{\lambda} = hc\overline{v}$$

$$h = \text{Planck's constant}$$
$$6.626 \times 10^{-34}\,\text{J.s}$$
$$c = \text{speed of light,}$$
$$3.00 \times 10^{8}\,\text{m.s}^{-1}$$

As the equation accompanying figure shows, radiation can be characterized by its frequency (v), its wavelength (λ), or its wavenumber ($\overline{v}$). The relationships between these quantities are:

$$v\left(s^{-1}\right) = \frac{c\left(\text{m·s}^{-1}\right)}{\lambda(\text{m})} \qquad \overline{v}\left(\text{cm}^{-1}\right) = \frac{1}{\lambda\ (m)} \times \frac{1\,\text{m}}{100\,\text{cm}}$$

Although the wavenumber (cm^{-1}) is not an S.I. unit, it is conventionally used to describe the transitions in infrared (IR) spectroscopy.

The unit of frequency, s^{-1} ("per second"), is known as a hertz (Hz). This unit is sometimes convenient for very low energy transitions, such as in nuclear magnetic resonance (NMR) spectroscopy. In general, an absorption spectrum is obtained by recording the amount of radiation absorbed by the sample as a function of the frequency or wavelength of the incident radiation. Each type of spectroscopy focuses upon a specific region of the electromagnetic spectrum. We will be primarily interested with infrared (IR) (4000 - 200 cm^{-1}) and nuclear magnetic resonance (NMR) (10 - 900 MHz) spectroscopies.

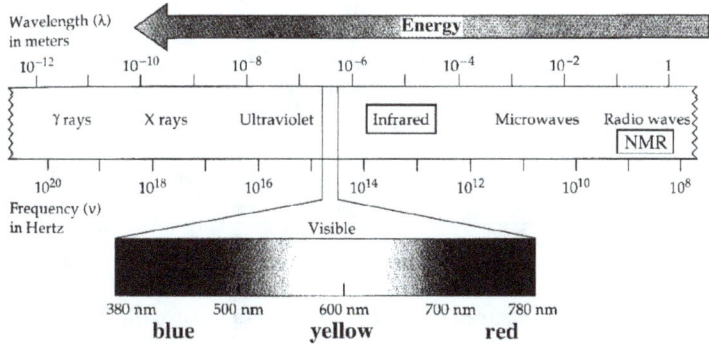

Figure: The electromagnetic spectrum.

There are following three types of Spectroscopy:

Infrared (IR) Spectroscopy

Infrared spectroscopy is used to study the vibrational motions of molecules. As shall be described shortly, it turns out that different motions among different groups of atoms cause the molecule to absorb different amounts of energy. Studying these transitions can sometimes allow us to determine what kinds of atoms are bonded or grouped in an unknown compound, which in turn gives clues as to the molecular structure.

IR spectroscopy $\Rightarrow$ identifying how certain atoms are bonded to each other or how they are grouped in a molecule

Theory1-4

Absorption of energy in the infrared region ($v- = 4000 - 200$ cm^{-1}) arises from changes in the vibrational energy of the molecules. There are two types of vibrations that cause absorptions in an IR spectrum. Stretching involves rhythmical displacement along the bond axis such that the interatomic distance alternately increases and decreases. Bending involves a change in bond angles between two bonds and an atom common to both.

(a) stretching   (b) bending

Figure: Stretching and Bending Vibrations.

For example, the borohydride anion (BH_4^-) has two vibrational modes that can be detected by IR spectroscopy.

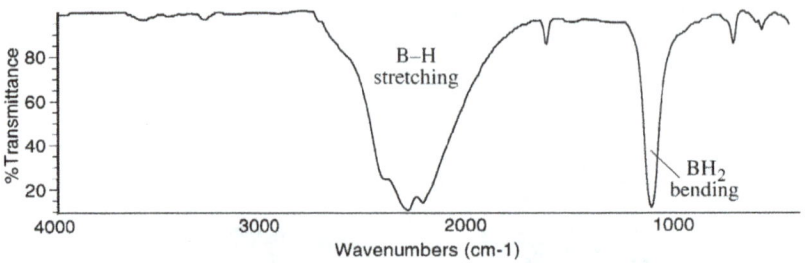

Figure: Infrared spectrum of NaBH4 (KBr pellet).

One important condition is that only those vibrations that produce a change in the electric dipole moment of the molecule will be observed in the infrared spectrum. For example, stretching vibrations in homonuclear diatomic molecules like O_2, N_2, and Br_2 do not produce a change in dipole moment and hence these molecules do not give rise to an IR spectrum. On the other hand, CO and IBr produce IR spectra because these molecules contain a permanent dipole moment that will change as the bond is stretched or compressed. CO_2, a linear molecule that does not have a permanent electric dipole, nevertheless produces an IR spectrum because the two C=O bonds can stretch in an asymmetric fashion and also bend to produce changes in the dipole moment. The symmetric stretch is not observed in the IR spectrum because it produces no change in the electric dipole moment, just as for homonuclear diatomics such as N_2.

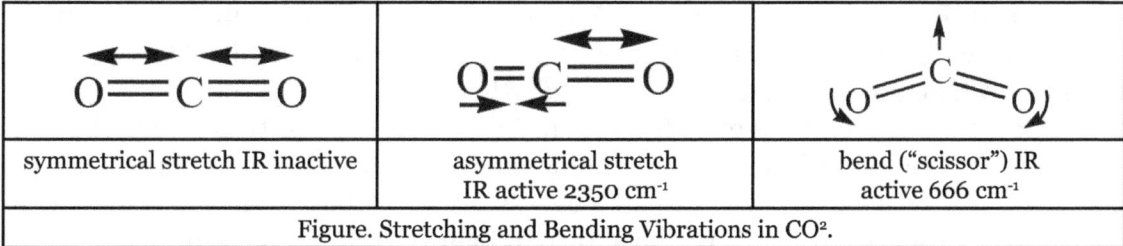

symmetrical stretch IR inactive	asymmetrical stretch IR active 2350 cm⁻¹	bend ("scissor") IR active 666 cm⁻¹
Figure. Stretching and Bending Vibrations in CO².		

An additional example is provided by the acetate ion, $CH_3CO_2^-$. In this case, the C–O vectors are not collinear, and both symmetrical and asymmetrical stretches are observed in the IR spectrum.

The stretching of a bond can be likened to the stretching of a spring, with the energy changes being detected by absorptions of IR radiation. While we will be treating IR spectroscopy from a qualitative standpoint, it is important to understand a few of the fundamental physical properties that determine the position of an absorption band in the IR spectrum. The following equation, derived from Hooke's law, describes the relationship between the frequency of oscillation ($\bar{v}$), the atomic masses (m_x and m_y), and the force constant of the bond (k).

$$\bar{v} = \frac{1}{2\pi c}\sqrt{\frac{k}{\mu}} \qquad \text{where} \qquad \mu = \frac{m_x m_y}{m_x + m_y}$$

The force constant, k, approximates the strength of the bond being stretched between two atoms. Thus, the stretching frequency of the C≡O triple bond (2143 cm-1) of carbon monoxide is higher than that of the C=O double bond in a ketone (1850 - 1650 cm⁻¹), which in turn is higher than that of a C–O single bond (1200 - 1000 cm⁻¹). Note that in the previous three examples, the atoms (and thus m_x and m_y) were kept constant and only k was varied.

However, an equally important component is the reduced mass, μ, which describes how the frequency will change as the masses of the two atoms change. This helps us understand why C–H stretches occur at higher frequency (3350 - 2850 cm⁻¹) than the C≡O triple bond (2143 cm⁻¹) of carbon monoxide even though the C≡O triple bond is a much stronger bond than the C–H single bond. The reduced mass of a C–H bond (μ = (12 × 1)/(12 + 1) = 0.92) is much smaller than that of the C≡O bond (μ = (12 × 16)/(12 + 16) = 6.86) and consequently leads to a larger frequency ($\bar{v}$) when inserted into the denominator of Hooke's law. Although Hooke's law demonstrates some of the fundamental features of IR spectroscopy, we will be interested primarily in qualitative applications. (Hint: although you will never be asked to perform a calculation involving Hooke's law, you should understand the factors that influence $\bar{v}$).

In theory, it is possible to predict the number of fundamental vibrations that will be observed in an IR spectrum.* In practice, IR spectra are more complicated that we might have expected. The infrared spectrum of gaseous BF3 provides an illustration of this. It turns out that trigonal planar molecules have four normal modes of vibration, three of which are IR active.1,2 (Why is the "breathing" (v_1) mode not IR active?) The B–F stretching (v_3) and out of-plane B–F bending modes (v_2) occur at approximately 1500 and 700 cm-1, respectively. The in-plane bending mode, while IR active, is too low in energy to be observed in the IR region shown in Figure 6 ($v4 = 481$ cm^{-1}).

One complication that is immediately apparent is the fact that two B–F stretches are observed at v_3 where we might have expected to see only one. This occurs because boron is composed of two isotopes: 10B (19.9% natural abundance) and 11B (80.1% natural abundance).

Therefore, a 10B–19F bond will have a different reduced mass ($\mu = (10 \times 19)/(10 + 19)$ = 6.55) and a different stretching frequency than a 11B–19F bond ($\mu = (11 \times 19)/(11 + 19) = 6.97$). The higher frequency band can be assigned to the 10B–19F stretch on the basis of its lower intensity (because of 10B's lower natural abundance), and this can be confirmed by applying Hooke's law:

$$\frac{\bar{v}\left(^{10}\text{B-F}\right)}{\bar{v}\left(^{11}\text{B-F}\right)} = \sqrt{\frac{\mu\left(^{11}\text{B-F}\right)}{\mu\left(^{10}\text{B-F}\right)}} = \sqrt{\frac{6.97}{6.55}} = 1.031 \quad \text{and} \quad \frac{\bar{v}\left(^{10}\text{B-F}\right)}{\bar{v}\left(^{11}\text{B-F}\right)} = \frac{1504\,\text{cm}^{-1}}{1453\,\text{cm}^{-1}} = 1.035$$

The two ratios agree to within ± 0.4 %, confirming our assignment. As another illustration of the effect of changing μ, the 10B–Cl and 11B–Cl stretches in BCl3 occur at 995 and 956 cm-1, respectively.

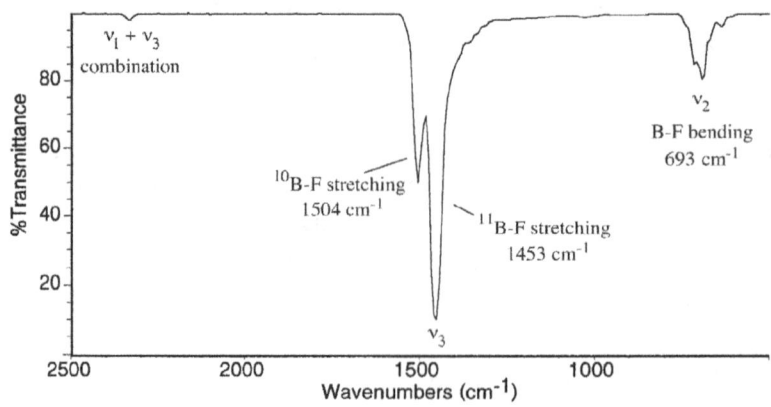

Figure. IR spectrum and fundamental vibrational modes of BF_3.

The second complication is the weak band at approximately 2330 cm⁻¹ that is not the result of a fundamental vibration. Most IR spectra will show many more than the number of bands predicted from a knowledge of the fundamental vibrational modes. Overtones occur when a vibrational mode is excited beyond the first excited state; the energy of the overtone band will therefore be higher than that of the fundamental and is often roughly equal to some multiple of the fundamental frequency. Combinations occur when more than one vibration is excited by the absorption of one photon. Combination bands occur at frequencies that are approximately equal to the sum of the two component vibrations. The band at ~2330 cm⁻¹ in the IR spectrum of BF3 has been assigned to a v1 + v3 combination.5,* Overtone and combination bands are seldom assigned in the qualitative analysis of IR spectra as their intensities are typically much weaker than those of fundamental vibrations. It is, nevertheless, important to remember their contribution to the appearance of IR spectra.

The previous example demonstrated that even the IR spectra of relatively small molecules can be quite complicated. Since we will usually be dealing with relatively large molecules, there are numerous possible stretching and bending motions and consequently a large number of infrared bands are usually observed in a particular spectrum. Assigning every band in the spectrum to a particular vibration is virtually impossible using qualitative techniques. Nevertheless, the IR spectrum of a molecule is very informative and can be used in the following ways:

(a) To identify the presence (or absence) of functional groups.

The vibrations of certain functional groups (e.g., C–H stretch, C=O stretch, P–H stretch, etc.) give rise to bands in well-defined frequency ranges regardless of the type of molecule that contains them. That is, their position is not greatly influenced by other atoms in the molecule. A listing of some group vibrations is provided in tables and figure. Functional groups within a molecule can be identified by comparing the bands observed in an IR spectrum with the frequency ranges in the correlation tables and figures. Remember that the ranges and intensities provided are guidelines, not hard and fast rules.

(b) As a fingerprint for molecule identification.

An unknown compound can be identified by matching its IR spectrum with that of a known compound. This type of analysis can be accomplished by a computer search of data banks of IR spectra of known compounds. The region below ~1500 cm-1 in an IR spectrum is particularly useful in this type of search, and is commonly referred to as the "fingerprint region". Because of the complexity of this part of the spectrum, you will be told when you should attempt assignments within the fingerprint region.

Table. IR Stretching Frequencies for Some Group Vibrations (cm⁻¹).

X–H	Stretch	Intensity[a]	X–H	Stretch	Intensity[a]
$-N\overset{R}{\underset{H}{\diagdown}}$	3500 - 3200	var	=C–H (sp^2)	3100 - 3000	w-m
			$-\overset{\vert}{\underset{\vert}{C}}$-H (sp^3)	3000 - 2840	m-s
–NH$_2$	3400 - 3300 (asym)[b]	var			
	3300 - 3250 (sym)[b]	var	S–H	2600 - 2550	w
–NH$_3^+$	3000 - 2800	m-s	B–H$_{terminal}$	2650 - 2250	var
	2800 - 2000[c]	m-w			
O–H	3600 - 3200	s, br	B–H$_{bridging}$	2200 - 1500	w-m
≡C–H (sp)	3350 - 3250	m-s	P–H	2450 - 2280	w-m

X≡Y	Stretch	Int.[a]	X=Y	Stretch	Int.[a]	X–Y	Stretch	Int.[a]
C≡O	2143		C=O	1850 - 1650	s	C–O	1300 - 900	s
RC≡CR'	2260 - 2190	w	C=C	1680 - 1630	w-m	B–N	1275 - 1075[e]	m-s
RC≡CH	2140 - 2100	w	C$_6$H$_5$–	1610 - 1400[d]	m	P–O	1100 - 900[e]	m-s
RC≡N	2260 - 2220	w-m	P=O	1300 - 1175[e]	m-s	B–P	650-600[e]	w-m
RN≡C	2175 - 2115	m-s	RCO$_2^-$ (asym)	1690 - 1560[b]	s			
			(sym)	1460 - 1310[b]	w-m			

[a] Intensities: s = strong, m = medium, w = weak, br = broad, var = varies. These intensities serve as a guide only; remember that frequency is much more diagnostic.

[b] Two bands are observed for this group; asym = asymmetric, sym = symmetric.

[c] In salts of primary amines, the 2800-2000 cm⁻¹ region consists of several combination bands.

[d] Refers to carbon-carbon stretching in the aromatic ring; weak overtone bands are commonly observed between 2000 - 1650 cm-1 when phenyl groups are present. e In practice, these functional groups span a larger frequency range than is indicated. The range provided reflects the types of compounds encountered in our courses.

Table. IR Bending Frequencies for Some Group Vibrations (cm⁻¹).[3,4]

Group	Bend	Intensity[a]
–NH2	1650 - 1550	var
–NH3$^+$	1600 - 1500	w
–CH3	1475 - 1350 (two bands)	var
–CH$_2$–	~1475 - 1450	m-s, w[b]
	1350 - 1150	

[a] Intensities: s = strong, m = medium, w = weak, br = broad, var = varies. These intensities serve as a guide only; remember that frequency is much more diagnostic.

[b] These bands are often obscured by stronger absorptions from other functional groups.

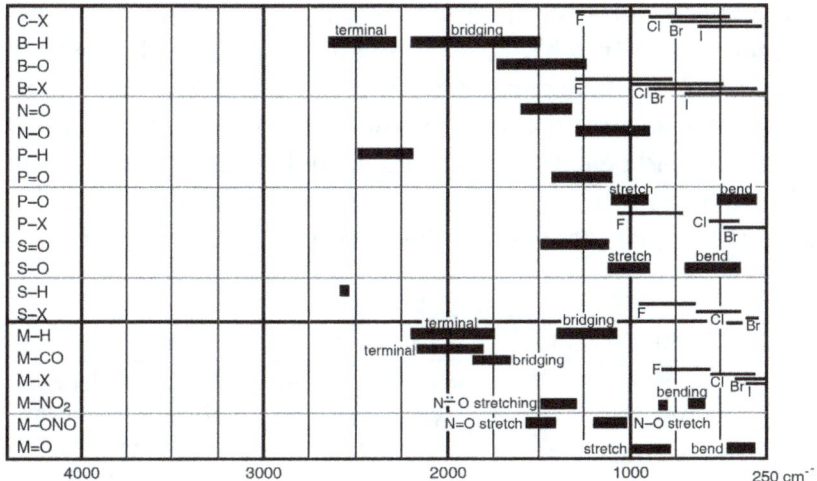

Analyzing IR Spectra and Reporting Results

The IR spectrum of tert-butylamine, (CH3)3CNH2, appears on page 10. Although bands appear in the region below ~1500 cm-1 (labeled "fingerprint region"), not all have been assigned to specific bond vibrations. This is because many bands (due to both stretching and bending vibrations) are commonly found below ~1500 cm-1, rendering exact assignment difficult. You will be told when you should make assignments within the fingerprint region. When interpreting an IR spectrum, focus your attention on bands that have reasonable intensity before considering weak bands; do not worry about assigning bands that barely register above the baseline.

Infrared spectra recorded in the laboratory should be labeled with your name, the date, the compound name or formula and the sample method (e.g., thin film). The major bands should be identified and labeled directly on the spectrum. The data obtained from the analysis of an infrared spectrum should also be summarized in a table. The table should report the band position, intensity, and proposed assignment. Band positions should be rounded to the nearest wavenumber and reported as ranges for broad peaks or groups of peaks (e.g., the C–H stretches in tert-butylamine).

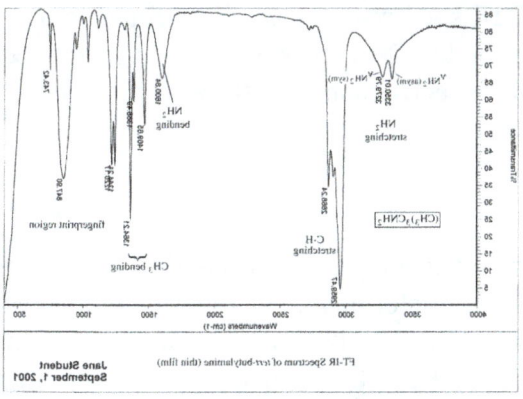

Band positions may be reported as a single wavenumber if the peak is sharp and distinct (e.g., each N–H stretch). Band intensities are described either in terms of %Transmittance (%T) or absorbance (A), and should be reported in relative terms using descriptors such as strong, medium, and weak. Bands are normally assumed to be relatively sharp; broad peaks should be identified in the summary. The assignment should specify the atoms that are vibrating and the vibrational mode (i.e., stretch or bend; additional descriptors such as "symmetric" or "asymmetric" may apply for groups).

Band Position (cm^{-1})	Intensity	Assignment
3350	weak weak strong	NH2 stretch (asym)
3280	weak (broad)	NH2 stretch (sym)
2970 - 2860	medium	C–H stretching NH2 bending
1600		CH3 bending
1470 - 1360		

Nuclear Magnetic Resonance (NMR) Spectroscopy

Nuclear Magnetic Resonance (NMR) spectroscopy takes advantage of the magnetic properties of certain nuclei and records the absorption of energy between quantized nuclear energy levels. In an NMR experiment, the spectrometer is tuned to the frequency of a particular nucleus and the spectrum reveals all such nuclei in the molecule being investigated. It is thus a very powerful technique, the closest analogy being a powerful microscope that allows the chemist to "see" the structure of molecules in solution. Actually, the NMR experiment does not directly show how all the atoms are connected. Accordingly, it is up to the chemist to take the information provided by NMR spectra to build a model of the molecule.

NMR spectroscopy $\Rightarrow$ establishing the number and connectivity of certain atoms in a molecule

The analysis of NMR spectra is very much like putting a puzzle together, and only when all of the pieces fit together will the structure of the molecule be known. Your problem solving skills will therefore be put to the test. In the rest of the lab course, the focus shall be on using NMR spectroscopy as a tool for elucidating the structures of the compounds prepared in the lab. Nevertheless, a certain appreciation of the theoretical background behind NMR spectroscopy must be in place before it can be successfully used as an analytical technique.

Theory1-4 (WWW)

NMR is possible owing to the magnetic properties of certain nuclei. In addition to charge and mass, which all nuclei have, various nuclei also possess a property called nuclear spin, which means that they behave as if they were spinning. Since nuclei have a charge, they generate a magnetic field with an associated magnetic moment.

There are useful empirical rules relating mass number, atomic number (Z) and nuclear spin quantum number (I):

Mass Number	Z	I
even	even	0
odd	even or odd	$^1/_2, ^3/_2, ^5/_2, ...$
even	odd	1, 2, 3, ...

Since NMR depends on the existence of a nuclear spin, nuclei with I = 0 have no NMR spectrum (e.g., 12C, 16O, 18O). From standpoint of generating NMR spectra, the most important class of nuclei are those with I = 1/2. Nuclei with I >1/2 (e.g., 11B, I = 3/2; 14N, I = 1) have quadrupole moments, a non-spherical distribution of nuclear charge, which results in broad absorption lines and makes observation of spectra more difficult. The quadrupole moment can even affect the lineshape of neighbouring nuclei. For example, resonances of protons bonded to nitrogen or boron atoms are generally broad in 1H NMR spectra. We shall thus be primarily concerned with nuclei where I = 1/2, but the effect that quadrupolar nuclei can have on the NMR spectra of I = 1/2 nuclei should be remembered. A listing of isotopes with I = 1/2 is provided in table.

Table: Natural abundances of isotopes with I = $^1/_2$.

Isotope	Natural Abundance (%)	Isotope	Natural Abundance (%)	Isotope	Natural Abundance (%)
^{1}H	100	^{107}Ag	51.35	^{129}Xe	26.44
^{13}C	1.108	^{109}Ag	48.65	^{169}Tm	100
^{15}N	0.365	^{111}Cd	12.75	^{183}W	14.4
^{19}F	100	^{113}Cd	12.26	^{187}Os	1.64
^{29}Si	4.71	^{115}Sn	0.34	^{195}Pt	33.8
^{31}P	100	^{117}Sn	7.57	^{199}Hg	16.84
^{57}Fe	2.17	^{119}Sn	8.58	^{203}Tl	29.50
^{77}Se	7.58	^{123}Te	0.87	^{205}Tl	70.50
^{89}Y	100	^{125}Te	6.99	^{207}Pb	21.7
^{103}Rh	100				

In an NMR experiment, the sample is placed in a strong magnetic field, Bo. Since the spins of the magnetic nuclei are quantized, they can have only certain well-defined values. If we have nuclei with $I = ^1/_2$ (e.g., ^{1}H, ^{31}P), the spins can orient only in two directions: either with $\left(m_I = +^1/_2, \alpha\right)$ or against $\left(m_I = -^1/_2, \beta\right)$ the applied field. NMR transitions are allowed for cases where ΔmI = ±1. There is an energy difference, ΔE, between the two states, and this is given by

$$\Delta E = h\nu = \frac{h}{2\pi}\gamma B_o \quad or \quad \nu = \frac{1}{2\pi}\gamma B_o$$

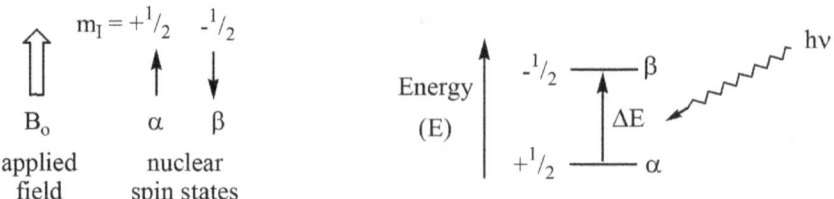

where h is Planck's constant, γ is the gyromagnetic ratio (a constant characteristic of each nucleus)*, and Bo is the applied magnetic field. When the energy of the incoming radiation matches (is in resonance with) the energy difference between the spin states, energy is absorbed and the nucleus is promoted from the lower +1 /2 to the higher -1 /2 spin state. Since the sign of m_I changes, this is sometimes referred to as a "spin flip". NMR transitions occur in the radio frequency (rf) range of the electromagnetic spectrum. The absorption of rf energy is electronically detected and is displayed as an NMR spectrum.

The above equation is very important since it shows that ΔE depends only on γ and Bo. The gyromagnetic ratio, γ, is an intrinsic property of the magnetic nucleus. Therefore, each type of nucleus has a distinct and characteristic value of γ. Accordingly, the NMR experiment must be tuned for a specific nucleus and one must record a different NMR spectrum for each NMR active nucleus of interest. Conversely, you do not have to worry about observing signals from different nuclei on the same NMR spectrum. In order to gather all NMR knowledge about a molecule such as PH_3, we would record two different NMR spectra - a 1 H NMR spectrum to observe the ^1H nuclei and a 31P NMR spectrum to observe the ^{31}P nucleus. We would not observe the ^{31}P nucleus in a ^1H NMR spectrum and vice-versa.

The above equation also reveals that ΔE is directly proportional to Bo, the external magnetic field. The higher the external field, the greater is the energy separation between the $\alpha \left(m_I = + \frac{1}{2} \right)$ and $\beta \left(m_I = - \frac{1}{2} \right)$ spin states. Recalling that E = hv, another way of saying this is that the resonance frequency of the nucleus increases with increasing Bo since if E increases, so does v.

This is shown in the following table.

Bo	Resonance Frequency (v, MHz)				
(tesla)‡	^1H	^{13}C	^{11}B	^{19}F	^{31}P
2.35	100	25.2	32.1	94.1	40.5
4.70	200	50.4	64.2	188.2	81.0

‡ a tesla is a unit describing magnetic field strength

Note that all I = 1 /2 nuclei behave according to the same theoretical principles - although ^1H NMR spectroscopy is the most commonly practiced, ^{19}F and ^{31}P NMR spectra are generated in exactly the same way as a ^1H NMR spectrum. The main difference between the different $I = 1/2$ nuclei is that the resonance frequency is changed when recording the spectrum.

Chemical Shift

A little reflection about the previous equation reveals that NMR would not really be useful at all to elucidate molecular structure if the relationship between γ and Bo was all there was to it. Indeed, since the resonance frequency is determined only by γ and Bo, all atoms of a given nucleus in a molecule (e.g., all ^1H nuclei) should resonate at the same frequency. If this were the case, the only thing NMR could tell us is whether a molecule contains NMR active nuclei (^1H, ^{31}P, ^{13}C, etc.). Fortunately, the frequency of the NMR absorptions of a given nucleus also depends on the chemical environment of the nucleus. The variation of the resonance frequency with chemical environment is termed the chemical shift, and herein lies the power of the NMR method.

The origin of the chemical shift can be traced to the electrons surrounding the nucleus, and the interaction of the electron cloud with the applied field, Bo. The reason for this is that circulating electrons also generate a magnetic field that orients itself in the opposite direction to the applied field.

The actual field (B_{local}) "felt" by a nucleus is thus less than B_o, and the ability of the electrons to alter the field felt at the nucleus can be expressed by σ, the shielding constant.

$$B_{local} = B_o(1-\sigma) \ or \ v_{local} = \frac{1}{2\pi}\gamma B_o(1-\sigma)$$

Nuclei are said to be shielded or deshielded depending on the presence or absence of electron density surrounding them. For example, the introduction of an electron withdrawing group (e.g., halogen, O, etc.) will reduce the electron density around a nucleus (deshielding; σ is small) and the resonance frequency will increase. Conversely, an electron donating substituent (e.g., CHx, SiHx) will cause increased shielding (σ is large) and lower the resonance frequency.

In reporting chemical shifts, one could use absolute field or absolute frequency, but this would be cumbersome and would result in the chemical shift being dependent upon the applied field. A simpler scale for chemical shifts has been devised. Chemical shifts (δ) are expressed in units of parts per million (ppm) of the spectrometer frequency with respect to a reference material whose position is arbitrarily assigned a value of 0.0 ppm.

$$\delta(\text{ppm}) = \frac{\nu_{sample}(Hz) - \nu_{reference}(Hz)}{\nu_{spectrometer}(Hz)} \times 10^6 = \frac{\Delta\nu(\text{spectrometer from reference in Hz})}{\nu_{spectrometer}(\text{in MHz})}$$

When expressed in such dimensionless units (δ in ppm), the chemical shifts are invariant of the frequency of the spectrometer and can be used as molecular parameters. For example, 1.0 ppm at 60 MHz is equal to a separation of 60 Hz, and at 200 MHz, 1.0 ppm equals 200 Hz. Thus, the same two resonances that are separated by 1 ppm at 60 MHz are still 1 ppm apart at 200 MHz, because δ = 60 Hz/60 MHz = 200 Hz/200 MHz = 1 ppm. Therefore, if the same sample is run at two different spectrometer frequencies, the chemical shifts of the resonances will be identical. Naturally, this statement is only true if the same reference material is used for each spectrum. Different references are used for different nuclei. The most widely accepted reference for ^{1}H and ^{13}C NMR is tetramethylsilane (Si(CH$_3$)$_4$ = TMS). For $_{11}$B NMR, F$_3$B•OEt$_2$ is commonly used, as are CFCl$_3$ for ^{19}F NMR and 85% H$_3$PO$_4$ for ^{31}P NMR spectroscopy.

In the past, NMR spectra were obtained by varying the applied field and measuring the chemical shift as a function of the field strength. This gave rise to the terminology of a downfield shift for nuclei that were deshielded (as they required a lower applied field to bring the nucleus into resonance)‡ and upfield shift for shielded nuclei. For example, one would say that a resonance at δ 8.0 ppm is downfield of one at δ 2.0 ppm, and conversely that the signal at δ 2.0 ppm was upfield of the signal at δ 8.0 ppm.

More modern NMR spectrometers generate spectra by varying the frequency, ν, while keeping the magnetic field strength, Bo, constant.§ Nevertheless, the upfield/downfield terminology remains in common use. Unfortunately, this results in the confusing situation that δ is positive in the downfield direction (to the left of the standard on spectra) where resonance frequencies are higher. Resonances that are upfield of the reference appear at lower frequencies and have negative δ values.

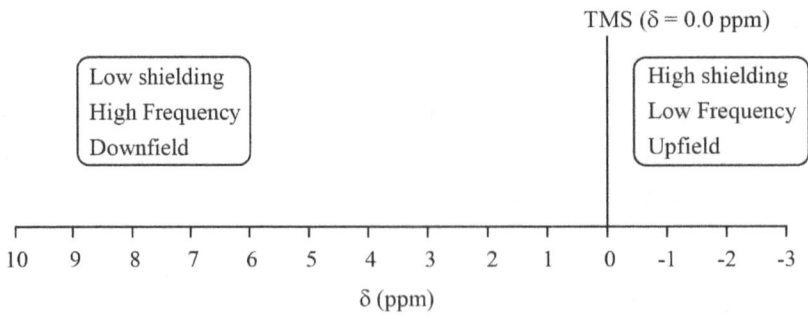

The concept of chemical shift is illustrated in figure below. As the hydrogens of methane are increasingly substituted by electron withdrawing chlorine atoms, the chemical shift of the remaining hydrogens shifts further downfield as the hydrogens become increasingly deshielded. Substitution of the methyl groups of tetramethylsilane (TMS) by chlorine has similar, but far less dramatic, results. In this case, the electron withdrawing chlorine atoms are separated from the hydrogens by carbon and silicon, resulting in less significant deshielding of the ^1H nuclei.

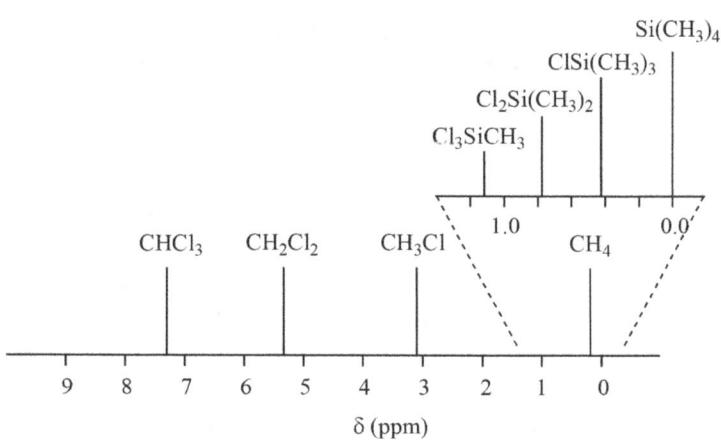

One important consequence of chemical shift is that each chemically different type of NMR-active nucleus in a molecule will give rise to its own signal in an NMR spectrum. Nuclei are thus referred to as chemically equivalent or chemically inequivalent in determining how many signals will be observed in an NMR spectrum. For example, both CH_3Cl and CH_2Cl_2 provided one resonance each in the ^1H NMR spectrum in figure above. From this, we can infer that the individual hydrogens in each of these molecules are chemically equivalent. From the viewpoint of chemical structure, the reason for this is that hydrogens are related by symmetry elements (reflection through a mirror plane or rotation about an axis) and are thus identical.

mirror in the plane of the
page renders $H^a = H^b$

three-fold rotation axis demonstrates
$H^a = H^b = H^c$

Sometimes, determining chemical equivalence or inequivalence is straightforward. It would not take very much to convince you that the methyl hydrogens in ethanol (CH_3CH_2OH) are different from the methylene hydrogens and that both of these are different than the hydroxyl hydrogen; we would thus anticipate three signals in the ^1H NMR spectrum. Upon further reflection though, why should the hydrogens of the methyl group all be equivalent? The answer is simple when it is recognized that methyl groups rotate freely and rapidly, with the result that each hydrogen experiences the same overall chemical shift as it completes one rotation, a situation analogous to CH_3Cl described above. Therefore, all methyl groups generally give rise to one signal in ^1H NMR spectra. This concept can generally be applied to analogous groups such as tert-butyl, $C(CH_3)_3$, trimethylsilyl, $Si(CH_3)_3$, and trifluoromethyl, CF_3 (in ^{19}F NMR spectra).

The most general method of determining whether nuclei are chemically equivalent to other nuclei in a molecule is to determine whether they are in the same environment, and whether one nucleus can be related to the other through a symmetry transformation such as rotation or reflection through a mirror plane. Some examples are provided below for illustration.

$$CH_3CH_2-O-CH_2CH_3$$

The CH_2 groups are equivalent and the CH_3 groups are equivalent. $\Rightarrow$ 2 signals in either the ^1H or ^{13}C NMR spectra

$$CH_3CH_2-O-CH_3$$

The CH_3 groups are inequivalent. $\Rightarrow$ 3 signals in either the ^{1} or ^{13}C NMR spectra

$$CH_3CH_2CH_2Cl$$

The CH_2 groups are inequivalent. $\Rightarrow$ 3 signals in either the ^1H or ^{13}C NMR spectra

$$ClCH_2CH_2CH_2CH_2Cl$$

There are two distinct sets of CH_2 groups. $\Rightarrow$ 2 signals in either the ^1H or ^{13}C NMR spectra

SF_6 is a highly symmetrical octahedral molecule $\Rightarrow$ 1 signal in the ^{19}F NMR spectrum

The axial (ax) and equatorial (eq) fluorines are chemically inequivalent $\Rightarrow$ 2 signals in the ^{19}F NMR spectrum

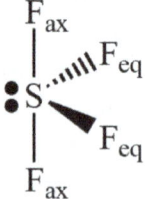

The apical fluorine is chemically distinct from the four fluorines in the square base $\Rightarrow$ 2 signals in the ^{19}F NMR spectrum

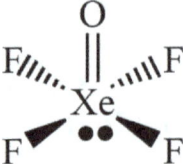

The four fluorine nuclei in the square base are chemically equivalent $\Rightarrow$ 1 signal in the ^{19}F NMR spectrum

Integration

The area under each NMR absorption peak can be electronically integrated to determine the relative number of nuclei responsible for each peak. The integral of each peak can be provided numerically, and is often accompanied by a line that represents the integration graphically. Intensities of signals can be compared within a particular NMR spectrum only. For example, ^{1}H intensities cannot be compared to those of ^{19}F or ^{31}P nuclei. It is important to note that the integration of a peak is a relative number and does not give the absolute number of nuclei that cause the signal. Thus, the 1 H NMR spectrum of $H_3C–SiH_3$ will show two peaks in a 1:1 ratio, as will the ^{1}H NMR spectrum of $(H_3C)_3C– Si(CH_3)_3$. This is simply because the ratios 3:3 = 9:9 = 1:1. Nonetheless, the integrated intensities of the signals in an NMR spectrum are a vital piece of the puzzle.

The concept of integration, and also that of chemical shift, is illustrated by figure. Determining integration ratios is an exercise in finding the greatest common divisor for the series of peaks (the largest whole number divisor that will produce a whole number ratio). In the above example, this value is either 1.4 cm or 9.9 integration units. It should be remembered that integration is a measurement that is subject to error; it is common for the error in integrated intensity to approach 5 - 10 %. The ratio of the integrated peak intensities is 1:3 = 3:9, allowing us to assign the resonance at δ 3.21 to the methyl group and that at δ 1.20 to the $(CH_3)_3C$ group. It is important to note that the hydrogens of the $(CH_3)_3C$ group are more shielded than the CH_3 group. This occurs because the CH_3 group is directly adjacent to the electron withdrawing oxygen, but the corresponding methyl protons in the $(CH_3)_3C$ group are separated from oxygen by a second intervening carbon center.

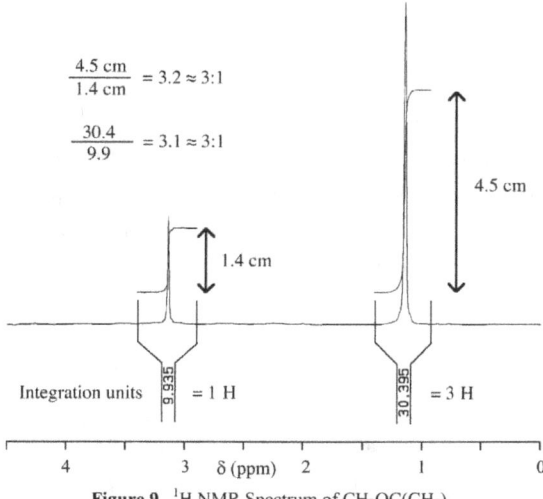

$$\frac{4.5 \text{ cm}}{1.4 \text{ cm}} = 3.2 \approx 3:1$$

$$\frac{30.4}{9.9} = 3.1 \approx 3:1$$

4.5 cm

1.4 cm

Integration units   9.935 = 1 H   30.395 = 3 H

4   3   δ (ppm)   2   1   0

Figure 9. ^{1}H NMR Spectrum of $CH_3OC(CH_3)_3$.

At this stage, we can begin to appreciate how NMR resembles a molecular microscope. For example, at one frequency we could "see" the various protons, while the carbons, fluorines, phosphorus, and even certain metal nuclei could be observed at other frequencies. Within one spectrum, we can make use of the position (chemical shift) and integrated intensity of the different signals to assign particular molecular fragments responsible for them, and to build up a model of the molecule. There is one more aspect of NMR that is extremely helpful in determining how to connect the parts together.

Spin-Spin Splitting (Coupling)

The appearance of a resonance may be very different when there are other neighbouring magnetic nuclei. The reason for this is that the nucleus under observation will interact with the magnetic spins of the different neighbouring nuclei.

The simplest case is that of two protons having significantly different chemical shifts (designated A and X). Considering chemical shift and integration only, we could represent the spectrum as:

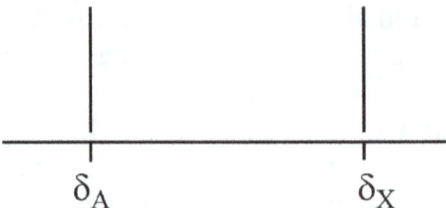

Both protons have a spin of $\frac{1}{2}$, and both can exist in the $+\frac{1}{2}$ and $-\frac{1}{2}$ spin states. Now, it turns out that the magnetic environment of H_A is slightly different when H_x is in the $+1/2$ state than when it is in the $-\frac{1}{2}$ state. This can be represented pictorially with arrows (pointing either up or down) representing the two spin states of H_x.

As a result, HA will split into two lines, each half the intensity of the unperturbed signal. Similarly, HA will influence HX which becomes a doublet also. The splitting, or coupling, is symmetrical about the unperturbed resonances δ_A and δ_x, and is described by the means of a coupling constant, JAX, which has units of Hz.

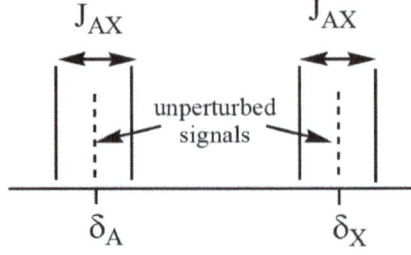

Note that the magnitude of JAX is identical at both signals - coupled nuclei must share the same coupling constant.

In a similar way, the resonance of a proton attached to phosphorus will be a doublet, since the phosphorus nucleus has I = $\frac{1}{2}$ and may be in the $+\frac{1}{2}$ or $-\frac{1}{2}$ state. However, the key distinction here is that we are dealing with two different nuclei, and thus two different NMR spectra. Each NMR spectrum (^{1}H and ^{31}P) will show one doublet with a JPH coupling constant that is identical in magnitude. Recall that we cannot "see" a ^{31}P nucleus in a ^{1}H NMR spectrum and vice-versa. Nonetheless, the splitting of the peaks into doublets in each spectrum tells us that the ^{1}H and ^{31}P nuclei are interacting.

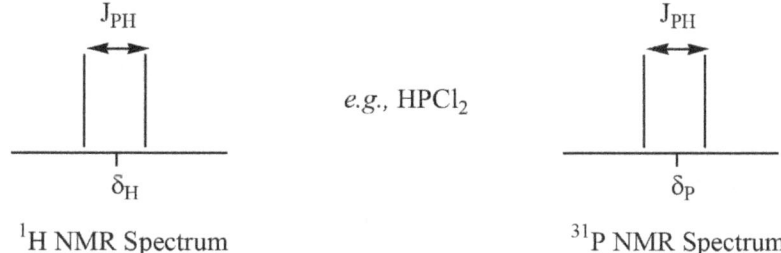

To review, the influence of the neighbouring spins is called spin-spin coupling and NMR peaks are split into multiplets as a result. The separation between the two peaks is called the coupling constant, J, which is expressed in Hz. Spin-spin coupling has the following characteristics:

- The magnitude of J measures how strongly the nuclear spins interact with each other.

- Coupling is normally a through-bond interaction, and is proportional to the product of the gyromagnetic ratios of the coupled nuclei. For example, $^1J_{CH}$ = 124 Hz for ^{1}H-^{13}C coupling in CH4, and $^1J_{SnH}$ = 1931 Hz for ^{119}Sn-H coupling in SnH4. This happens because $\gamma(^{119}Sn)$ is much larger than $\gamma(^{13}C)$.

- Since coupling occurs through chemical bonds, the magnitude of J normally falls off rapidly as the number of intervening bonds increases. e.g., JPH ~ 700; ^{2}JPH ~ 20 Hz in

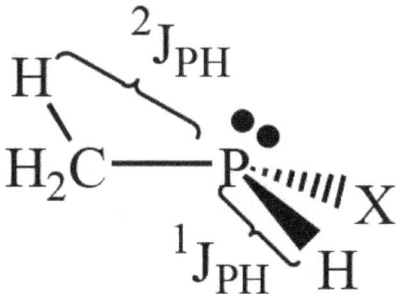

Coupling constants are thus labeled to show the types of nuclei and the number of bonds separating the nuclei that give rise to spin-spin splitting.

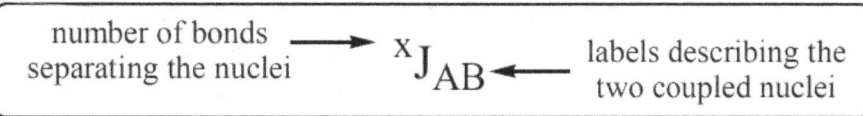

- Since spin-spin coupling is a through-bond interaction, it is sensitive to the orientation of the bonds between two interacting nuclei. This is particularly important for two-bond coupling constants. The influence of the orientation of the two coupled nuclei can occasionally render $^2J < {^3}J$. For example,

$$^2J_{H^aH^b} << {^3}J_{H^aH^c} < {^3}J_{H^bH^c}$$

^1J is not affected by the orientation of the coupled nuclei, so it is generally true that $^1J >> {^2}J$ or ^3J, but it is not always true that $^2J > {^3}J$.

- Spin-spin interactions are independent of the strength of the applied field. The spacing (in Hz) between lines at two different field strengths will be the same if it is due to coupling, but will be proportional to the field strength if it is due to a difference in chemical shift.

Table. Typical Coupling Constant Ranges (in Hz).

Coupled Nuclei (AB in ^XJAB)

x	HH	CH	PH^b	PC^b
1	–	115 - 250	630 - 710	120 - 180
2^a	2 - 30	5 - 60	7 - 13	5 - 40
3	2 - 17	2 - 20	6 - 11	5 - 11
4	–	–	0 - 1	–

^{a} Two bond couplings are particularly sensitive to the geometrical arrangement of the nuclei, which in some cases may render $^2J_{AB} < {^3}J_{AB}$. ^{b} Restricted to acyclic compounds.

Cases involving more than two nuclei with $I = \frac{1}{2}$ are direct extensions of the above. However, because there are more nuclear spins interacting, the pattern of lines observed in the NMR spectrum becomes more complicated. For example, let's consider the ^1H NMR spectrum of the HF_2 - anion (i.e., [F--H--F]-). We are observing the ^1H nucleus, but it is coupled to two chemically equivalent ^{19}F ($I = \frac{1}{2}$) nuclei. There are four ways that we can arrange the nuclear spins of the two fluorine nuclei, but only three different energy states are created, as is explained below:

intensity ratio: 1   2   1

Extending what we learned about the generation of a doublet, we can clearly see that the ^1H environment where both ^{19}F spins are "up" is different from that where both ^{19}F spins are "down". However, we can also arrange things so that one ^{19}F spin is "up" and the other is "down". The latter case is degenerate; that is, there is more than one way of accomplishing an "up/down" arrangement of nuclei, but each "up/down" arrangement has the same energy. As a result, a pattern of three peaks (or triplet) with an intensity pattern of 1:2:1 is generated as shown above. It is important to note that each line in the triplet is separated by the same $^1J_{HF}$ coupling constant. As we would expect, the ^{19}F NMR spectrum of HF_2 - would show a doublet because the fluorine nuclei are chemically equivalent and couple to one ^1H nucleus.

Another way of looking at this is to begin with a singlet for the ^1H nucleus and then couple each ^{19}F nucleus one step at a time. The coupling of the first ^{19}F nucleus generates a doublet. When each line in this doublet is split again into a doublet, they overlap identically at the center of the signal, generating a single line of intensity two relative to each outer line of intensity one:

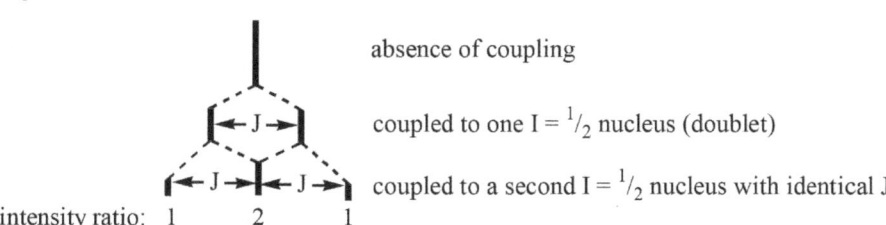

When a similar exercise is undertaken for the ^{31}P NMR spectrum of PF_3, * the nuclear spins of the three equivalent ^{19}F nuclei can be arranged in four ways to generate a quartet

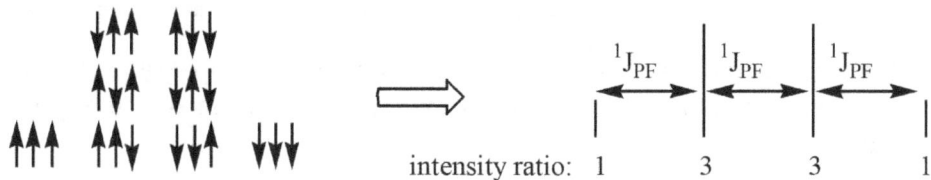

or we can split a singlet into doublets three times to accomplish the same transformation:

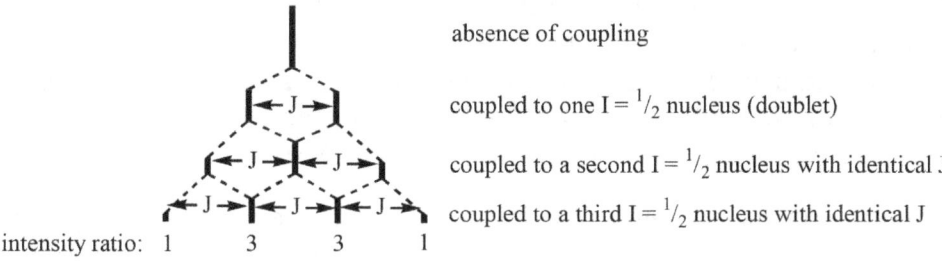

In this case, when each line at the triplet stage is split again into doublets, the intensity of the overlapping peaks is not identical; a signal of relative intensity two (from the middle peak) overlaps with a signal of intensity one (from the outer peak) to create a peak of intensity three.

Fortunately, the pattern of peaks generated by the interaction of $I = \frac{1}{2}$ nuclei can be easily generated by remembering that one nucleus is split by (n) equivalent nuclei into (n+1) peaks, each separated by the coupling constant, $^{x}J_{AB}$. The number of peaks is referred to as the multiplicity. The intensity pattern is a direct consequence of the number of combinations of the various nuclear spins that are possible and is described by a series of binomial coefficients. In practice, it is easiest to determine the intensity pattern by use of a mnemonic device such as Pascal's triangle.

n	$n+1$	Intensity	Multiplicity	Pattern	Example
0	1	1	singlet (s)		CH_4
1	2	1 : 1	doublet (d)		$(CH_3)_2CHCl$
2	3	1 : 2 : 1	triplet (t)		CH_3CH_2Cl
3	4	1 : 3 : 3 : 1	quartet (q)		CH_3CH_2Cl
4	5	1 : 4 : 6 : 4 : 1	quintet		$^{29}SiF_4$
5	6	1 : 5 : 10 : 10 : 5 : 1	sextet		PF_5^{*}
6	7	1 : 6 : 15 : 20 : 15 : 6 : 1	septet		$(CH_3)_2CHCl$

etc.

* An example of a case where the five fluorine nuclei are rendered equivalent by chemical exchange

The phenomenon of spin-spin coupling and its effect on the appearance and interpretation of NMR spectra is best described by example, several of which appear on the following pages.

Analyzing NMR Spectra and Reporting Results

NMR spectra contain a wealth of information and must be analyzed in a methodical way. Much like a jig-saw puzzle, all of the pieces (i.e., chemical shift, integration, multiplicity, and coupling constants) must fit together properly. As with a puzzle, you may find that your initial conclusion is incorrect because several "pieces" are out of place. It is important to approach the problem in a creative way and investigate alternate solutions. The most straightforward method for analyzing NMR spectra is:

1) identify signals by chemical shift and determine their relative integration

2) identify the multiplicity of the peaks and calculate coupling constants.

Many students are tempted to "leap in" and attempt to analyze coupling patterns first, but the coupling pattern may not correlate if the integration ratio of the coupled multiplets has not already been deduced. Above all else, remember to double-check that the assignments make sense. It is often a good practice to use your results to generate a simple stick-diagram of the NMR spectrum (e.g., Examples 1 and 7 on the following pages). If the stick-diagram matches the actual spectrum exactly, then you have correctly analyzed the NMR spectrum.

Clear communication of the results of interpretation of NMR spectra is vital. You should therefore label your spectra with pertinent information (such as your name, peak assignments, how the integration was derived, identification of coupling constants using appropriate $^x J_{AB}$ notation). Calculations and explanations of complex coupling patterns (e.g., Examples 6 and 7) should be shown directly on the spectrum whenever possible. NMR spectra (generally provided as handouts by the T.A.) should be taped or stapled into your lab notebook to form part of your lab report (or included in an Appendix in the case of the Formal Lab Report).

The data extracted from NMR spectra should also be summarized in a table. The objective is for your summary to be brief, yet comprehensive enough so that the spectrum could be simulated from the information provided in the table. It is also important to briefly explain your assignments so that your reader understands how you arrived at your conclusion. In the case of complicated coupling patterns, an explanation to clearly show the source of each contributing coupling constant (such as sketch of a "coupling tree") is usually appropriate.

Chemical shifts should generally be reported to two decimal places. Multiplicities may be written out (e.g., "triplet") or expressed in terms of common abbreviations (e.g., "t"). Coupling constants are commonly reported as whole numbers, but may be expressed to one decimal place if the spectrum is of sufficiently high resolution. If peaks are picked in ppm, you should show how you calculated the coupling constant(s). The coupling constants should be properly labeled (i.e., $^x J_{AB}$) to show the nuclei that are coupled; if there is more than one NMR active isotope for a nucleus (e.g., $^{117}Sn/^{119}Sn$), it should be clearly defined which is involved in the coupling interaction you are describing. Integration ratios are given in terms of whole numbers of nuclei, and you should demonstrate to your reader how you arrived at the ratio (i.e., did you measure the height of the integration line or were you relying on the integration unit values provided?).

For example, the data from the 1H NMR spectrum of $B(OCH_2CH_2)_3N$ would be summarized as:

Chemical Shift δ(ppm)	Integration	Multiplicity	Coupling Constant J (Hz)	Assignment
3.05	10.2 units = 2 H (or 88 mm = 2 H)	triplet (or t)	$^3 J_{HH}$ = 5.7 Hz	NCH2
3.89	10.0 units = 2 H (or 86 mm = 2 H)	triplet (or t)	$^3 J_{HH}$ = 5.7 Hz	OCH2

At least one sample calculation should be provided for full credit; e.g.,

$$^3J_{HH} = \frac{\delta \dfrac{(3.914 - 3.857)}{2} \times + (200 \times 10^6\ Hz)}{10^6} = 5.7\ Hz$$

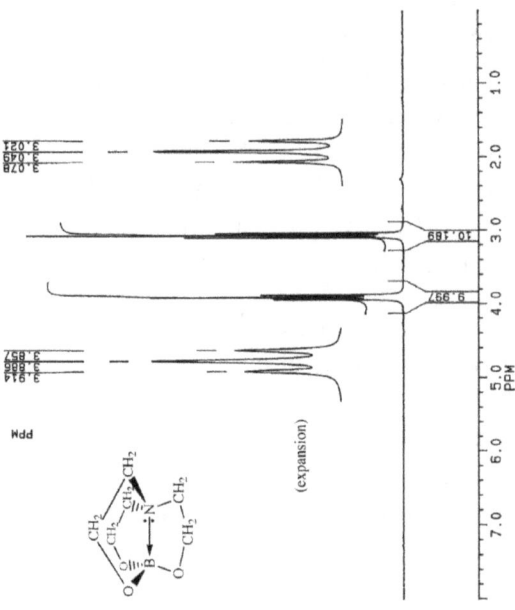

Example: ^{1}H NMR spectrum of $(CH_3CH_2O)_4Si$.

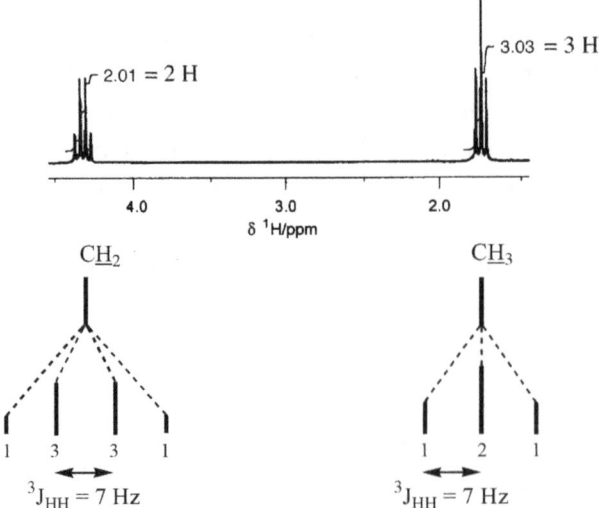

1. On the basis of chemical shift and integration, a CH_2 signal of intensity two appears downfield of a CH_3 signal of intensity three.

2. The CH_3 signal will be split into a triplet by interaction with the two equivalent

methylene protons (n = 2 and thus n+1 = 3). The CH_2 signal is split into quartet by the three equivalent CH_3 protons (n = 3 and thus n+1 = 4).

3. The spacing is ^3JHH = 7 Hz, and is the same in both regions.

4. The relative peak heights in the methyl triplet will be 1 : 2 : 1 and will be 1 : 3 : 3 : 1 for the methylene quartet. Recalling the overall integration, the methyl absorption must be $\frac{3}{2}$ as intense as methylene absorption as the total signal intensity is proportional to the number of nuclei; the integration ratio is 3.03÷2.01 ~ $\frac{3}{2}$.

Example: ^1H and ^{19}F NMR spectra of CHFCl2. Both spectra appear as doublets with an equal coupling constant, $^2J_{HF}$ = 51 Hz.

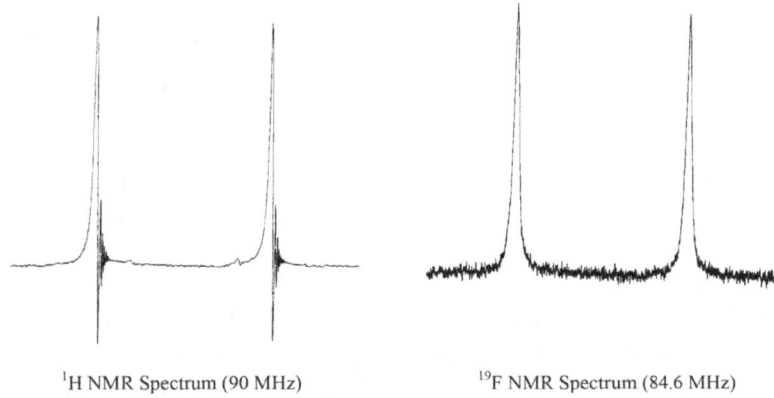

^{1}H NMR Spectrum (90 MHz) ^{19}F NMR Spectrum (84.6 MHz)

Example: ^{19}F and ^{31}P NMR spectra of an ion of the type $[P_xF_y]$. How can we use the two spectra to identify the unknown ion?

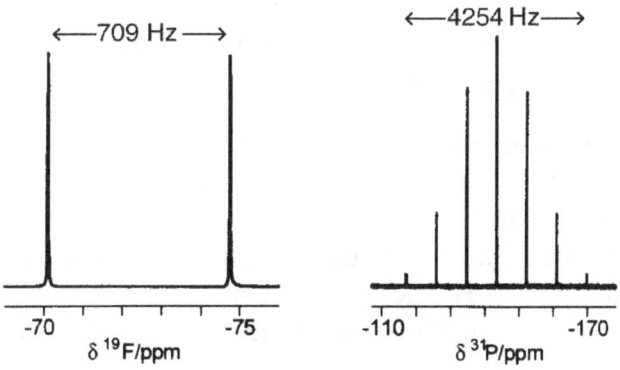

1. The ^{19}F NMR spectrum shows a doublet with a large coupling constant, so we can safely assume that we have an unknown number of equivalent fluorine atoms bonded to a single phosphorus (n+1 = 2 ⇒ n = 1).

2. The ^{31}P NMR spectrum shows seven lines with an intensity pattern similar to what we would expect for a binomial distribution. There are seven lines, so n+1

$= 7 \Rightarrow n = 6$. The unknown ion is therefore [PF6] - . We can confirm that the signal pattern is due to coupling by calculating $^1J_{PF} = (4254 \div 6)$ Hz $= 709$ Hz, which is the same as for the doublet in the ^{19}F NMR spectrum.

Example: If there are more than two types of NMR active nucleus in a compound, the pattern can be explained by the method of successive splittings. In this example, we shall consider an AMX system where $J_{AM} > J_{MX} > J_{AX}$.

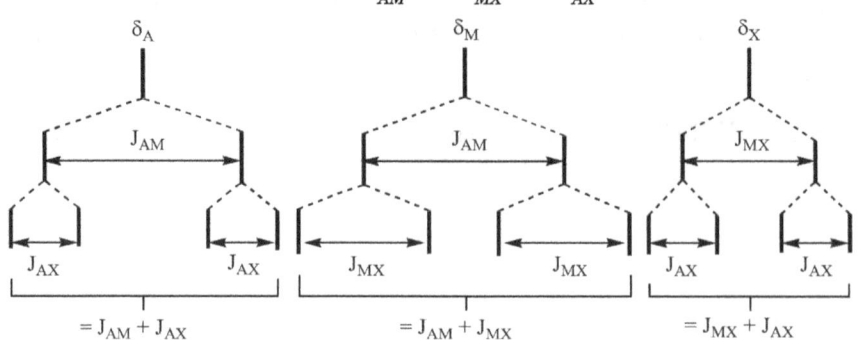

1. We begin with three equal intensity lines for the three nuclei of different chemical shift.

2. We split each signal successively, beginning with whichever coupling constant we like. Then, each resulting line is split again by the second (different) coupling constant. Note: the coupling constant can only affect nuclei with which it is associated (e.g., the signal pattern for nucleus A does not include J_{MX}).

The result is a total of twelve lines: four for A, four for M and four for X. Each pattern of four lines is referred to as a doublet of doublets. It is important to note that the distance separating the two outermost lines of each signal pattern is equal to the sum of the coupling constants that generate it.

If all of the nuclei were of the same type, for instance all were hydrogens in a compound such as $Cl_2CH-CHBr-CHI_2$, the entire pattern would appear in the same NMR spectrum (^1H in our example). However, if the nuclei each belonged to a different element, a pattern of four lines would appear in each of the three spectra. For instance, HPFCl would give ^1H, ^{19}F, and 31P NMR spectra, each appearing as doublets of doublets. Example 6 provides another illustration of an AMX spin system involving ^1H, ^{19}F, and ^{31}P NMR spectra.

Hint: While it does not matter how you generate the "tree diagrams", it is often easiest if you start with the largest coupling constant first followed by the smaller one(s).

Example: As opposed to the previous simple diagram where it was clear that we were dealing with three doublets of doublets, if the magnitudes of the coupling constants are similar, accidental overlap can sometimes cause the signal to appear differently than would be expected. Consider an AMX system where the signal at δ_X is being observed and the magnitude of J_{MX} is varied while J_{AX} is held constant:

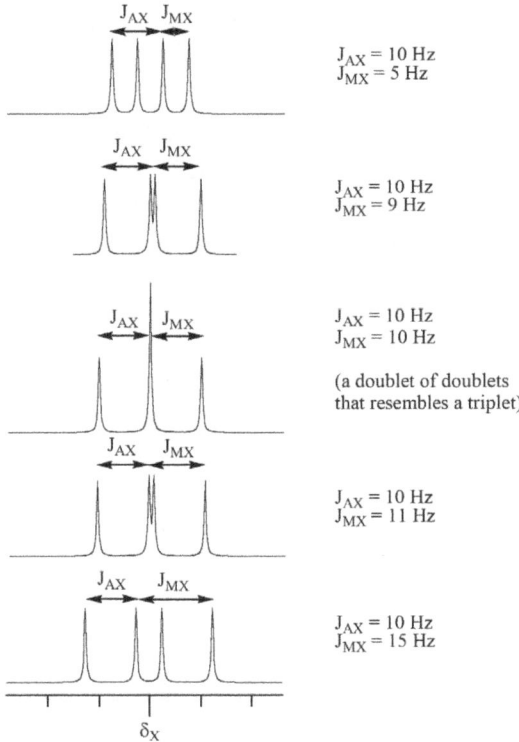

Example: The ^1H, ^{19}F, and ^{31}P NMR spectra of HOP(O)FH[9] form a simple AMX spin system as described in Example 4. In HOP(O)FH, $^1J_{PH} = 780$ Hz, $^1J_{PF} = 1030$ Hz, and $^2J_{FH} = 115$ Hz. Coupling is not observed between phosphorus and the proton of the hydroxyl group.*

^1H NMR Spectrum (recorded at 200 MHz):

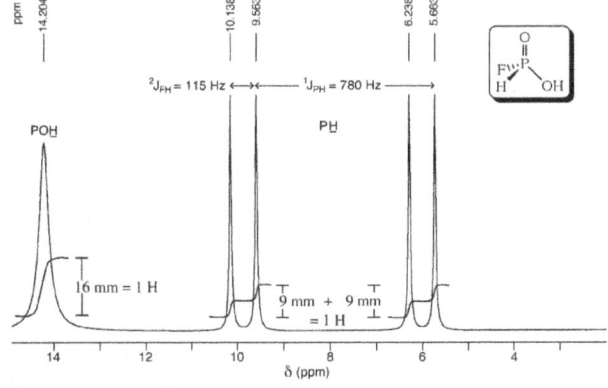

Recall that we can calculate coupling constants when the chemical shift of each peak in the splitting pattern is known. e.g.,

$$^1J_{PH} = \frac{\delta(9.563 - 5.663) \times (200 \times 10^6 \ Hz)}{10^6} = 780 \ Hz$$

$$^2J_{FH} = \frac{\delta(10.138 - 9.563) \times (200 \times 10^6 \ Hz)}{10^6} = 115 \ Hz$$

Convince yourself that the calculation would work just as well calculating 2 JFH from the peaks at δ 6.238 and 5.663 and 1 JPH from the peaks at δ 10.138 and 6.238. When it is possible to calculate the coupling constant from more than one set of peaks, the average should be reported. Note that the chemical shift of the PH proton is (δ 10.138 + 5.663)÷2 = δ 7.90.

¹⁹F NMR Spectrum:

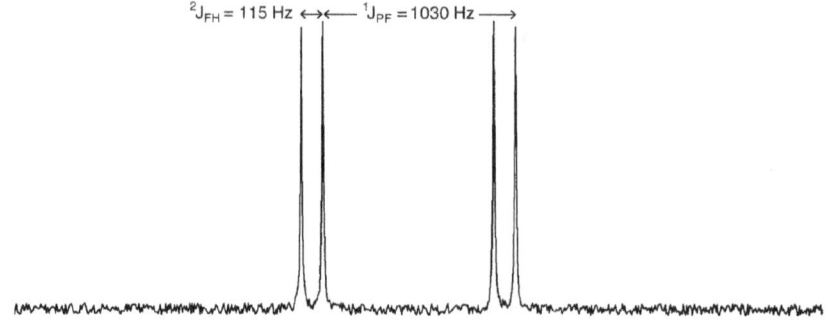

It is important to remember that each coupling constant will have the same magnitude (in Hz) in each spectrum in which it appears. Thus, $^2J_{FH}$ has the same magnitude in both the ¹H and ¹⁹F NMR spectra, as does $^1J_{PF}$ in the 19F and ³¹P NMR spectra and 1J$_{PH}$ in the ¹H and ³¹P NMR spectra.

³¹P NMR Spectrum:

Example: ¹H, ¹⁹F, and ³¹P NMR spectra of HP(O)F$_2$. ⁹In this example, $^1J_{PH}$ = 880 Hz, $^1J_{PF}$ = 1110 Hz, and $^2J_{FH}$ = 115 Hz.

¹H NMR Spectrum (doublet of triplets):

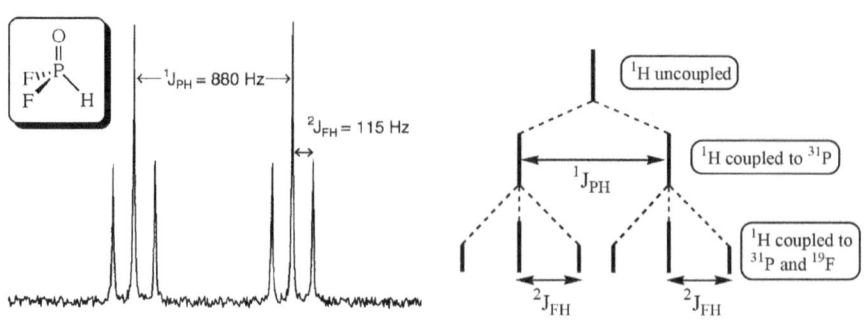

¹⁹F NMR Spectrum: (doublet of doublets)

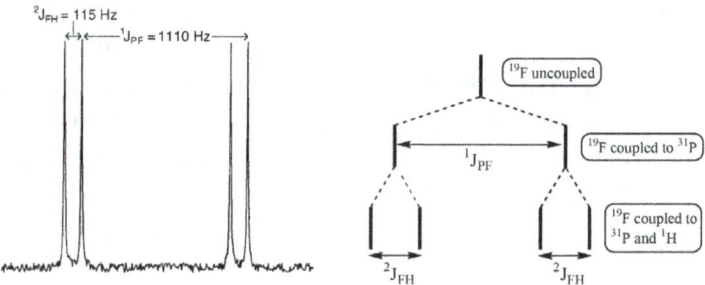

³¹P NMR Spectrum (triplet of doublets):

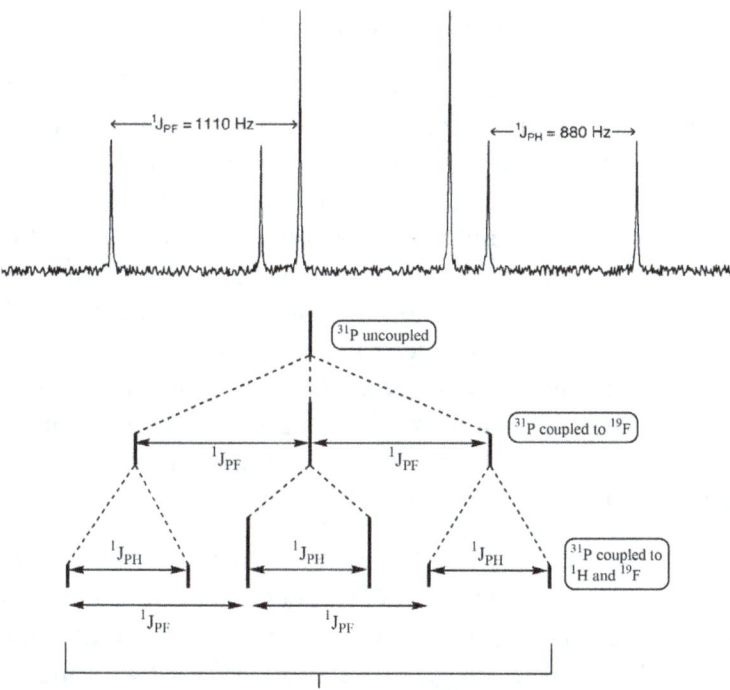

Complex coupling patterns such as these are named by describing the individual multiplicities in order of decreasing magnitude of J. Thus, ^{1}H NMR spectrum in this example is properly described as a doublet of triplets ($^1J_{PH}$ = 880 Hz, $^2J_{FH}$ = 115 Hz) or "dt". On the other hand, the ^{31}P NMR spectrum is a triplet of doublets ($^1J_{PF}$ = 1110 Hz, $^1J_{PH}$ = 880 Hz) or "td".

However, when drawing out the "coupling tree" to understand the pattern, it does not matter in which order we apply the couplings – the same result will be obtained. To convince yourself of this fact, draw the pattern for the 31P NMR spectrum to scale by applying the doublet splitting ($^1J_{PH}$) first and then the triplet splitting ($^1J_{PF}$).

A Few Advanced Topics in Spin-Spin Coupling:

(a) Decoupling: The examples discussed in Experiment 1 have demonstrated that spin-spin coupling information can be invaluable in assigning NMR spectra and predicting molecular structures. However, cases may arise where the large number of NMR active nuclei in a sample makes the interpretation of spectra quite difficult. Sophisticated methods exist to remove spin-spin interactions of certain nuclei when recording NMR spectra; decoupling is most frequently carried out when the sample contains a large number of ^1H nuclei. For instance, ^{31}P NMR spectra of organo-phosphorus compounds are frequently (but not always) recorded with ^1H nuclei decoupled. The result is designated as a $^{31}P\{^1H\}$ NMR spectrum where the "$\{^1H\}$" notation denotes proton decoupling.

The benefit of this technique can be seen immediately for a compound such as triethylphosphine, $P(CH_2CH_3)_3$, where we would expect $^2J_{PH}$ and $^3J_{PH}$ couplings to be visible. The ^{31}P NMR spectrum would be predicted to be a septet of decets (i.e., 70 lines in all); the $^{31}P\{^1H\}$ NMR spectrum, on the other hand, would be a singlet.

(b) Low abundance nuclei: The nuclei that we have considered thus far $\left(^1H,\ ^{19}F,\ ^{31}P\right)$ all occur in 100% abundance in nature. However, as detailed in table (Experiment 1), there are many $I = \frac{1}{2}$ nuclei that are not the only naturally occurring isotope of an element (^{13}C, 1.1% natural abundance, is a good example and will be discussed in Example). Cases such as this are sometimes referred to as being "spin dilute" to describe the effect that the low natural abundance has on recording NMR spectra. For example, ^{29}Si accounts for only 4.7% of naturally occurring silicon. Therefore, when recording a ^{29}Si NMR spectrum, only 4.7% of the silicon in the sample will generate an NMR signal. From an instrumental standpoint, this means that the signal-to-noise ratio will be quite poor and a more concentrated sample and longer spectral acquisition time may be necessary to compensate. The effect of spin-dilute systems on the appearance of coupling patterns is described in the following example.

Example: The ^{29}Si NMR spectrum[6] of SiF_4 shows a 1:4:6:4:1 quintet as we would anticipate. However, the corresponding ^{19}F NMR spectrum6 does not appear as a simple doublet. The spectrum is composed of a singlet (since 95.3% of the silicon is not NMR active) overlapped with a doublet ($^1J_{29SiF}$ = 178 Hz) for the 4.7% of the silicon that is ^{29}Si with $I = \frac{1}{2}$. The peaks due to the ^{29}Si-^{19} F coupling interaction in the ^{19}F NMR spectrum of SiF_4 are often described as satellites because of their low intensity.

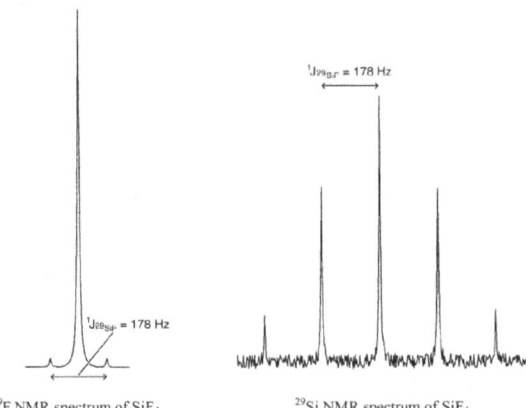

^{19}F NMR spectrum of SiF_4 ^{29}Si NMR spectrum of SiF_4

Example: Carbon is found in the greatest number of molecules save hydrogen. The bulk of naturally occurring carbon (12C, 98.9%) does not have a nuclear spin. Fortunately, the ^{13}C nucleus is magnetically active (I = $\frac{1}{2}$) but is present in only 1.1% natural abundance. Spectral acquisition is therefore challenging, but from the point of view of analysis, ^{13}C NMR is analogous to the other types of NMR of I = $\frac{1}{2}$ nuclei with one key exception. Because only 1 out of 100 carbon nuclei are detectable by NMR, the statistical chance of finding two neighbouring ^{13}C nuclei is very remote and 13C-13C spin-spin coupling is consequently not observed in ^{13}C NMR spectra. However, coupling to other I = $\frac{1}{2}$ nuclei, such as ^{1}H, ^{19}F, or ^{31}P, is observed in exactly the manner we would expect.

Since the vast majority of carbon nuclei contain attached protons and $^{1}J_{CH}$ is large (115 - 250 Hz), ^{13}C NMR spectra are frequently collected in the proton decoupled mode. Thus, in the absence of any other spin $\frac{1}{2}$ nuclei such as ^{19}F or ^{31}P, the ^{13}C{^{1}H} NMR resonances appear as singlets.

The chemical shifts of carbon nuclei generally mimic those of the protons to which they are attached, with the exception that the magnitude of δ is much greater in ^{13}C NMR spectra. One final note is that the integration of different types of ^{13}C nuclei is generally not very reliable, and thus the spectra are seldom integrated. However, the integration of similar types of carbons may be compared; an example is the ^{13C} NMR of transition metal carbonyl compounds.

The ^{13}C{^{1}H} and ^{13}C NMR spectra of diethyl ether appear below. Note the differences between the two types of spectra, especially the complication that $^{2}J_{CH}$ coupling is observed in the expanded regions of the ^{13}C NMR spectrum.

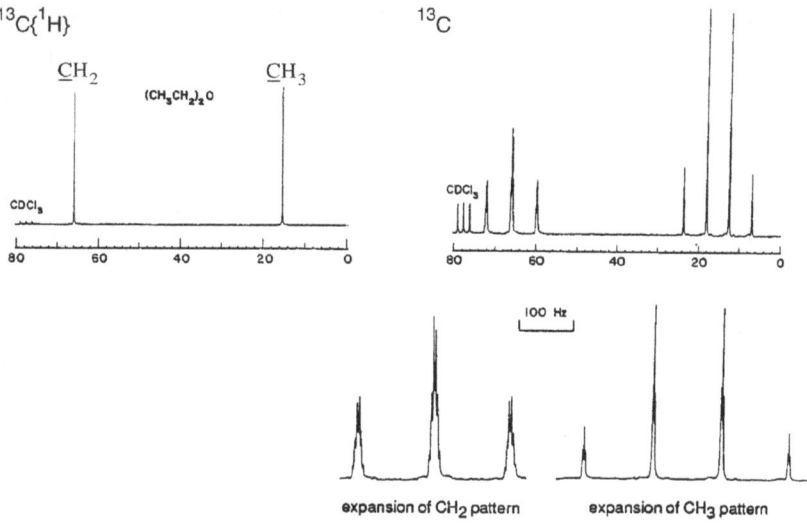

Also note that the carbon resonance of the NMR solvent, $CDCl_3$, appears as three equal intensity lines. This occurs because the I = $\frac{1}{2}$ ^{13}C nucleus couples to the I = 1 deuterium (D = ^{2}H) nucleus.

(c) Coupling to nuclei where $I > \frac{1}{2}$: In principle, it is no more difficult to record the NMR spectra of nuclei with greater nuclear spin than $\frac{1}{2}$. In practice, however, nuclei with $I > \frac{1}{2}$ possess a quadrupole moment, a non-spherical distribution of nuclear charge, that results in broad absorption lines and makes observation of spectra more difficult. Generally speaking, the larger the influence of the quadrupole moment, the broader is the NMR spectrum of the quadrupolar nucleus, and the greater is the broadening influence exerted on neighbouring nuclei. $^{10}B \; (I = 3)$, $^{11}B \left(I = \frac{3}{2}\right)$, and $^{14}N \; (I = 1)$ are some of the more commonly encountered quadrupolar nuclei.

When NMR spectra of quadrupolar nuclei can be recorded, spin-spin coupling interactions can occasionally be resolved. Although the lineshape is usually much broader than for the NMR spectra of $I = \frac{1}{2}$ nuclei, when the magnitude of coupling constants is large enough, coupling interactions to attached $I = \frac{1}{2}$ nuclei can be observed in the expected patterns from the n + 1 rule.

Most often, the primary influence associated with having $I = \frac{1}{2}$ nuclei attached to quadrupolar nuclei is that their signals are broad, sometimes to the point of not being observable. On the rare occasions where the influence of the quadrupolar nucleus is relatively small, it is possible to observe coupling to $I > \frac{1}{2}$ nuclei in the NMR spectra of $I = \frac{1}{2}$ nuclei (however, do note that in most cases, the broadening influence of the quadrupolar nucleus renders this impossible). Because of their larger magnetic moment, nuclei with $I > \frac{1}{2}$ have more spin states than the two associated with spin $\frac{1}{2}$ nuclei ($m_I = +\frac{1}{2}, -\frac{1}{2}$). The spin states are quantized, and transitions are only allowed when $\Delta m_I = \pm 1$. For example, ^2H (deuterium, also recognized by the symbol D) has I = 1, and three spin states $(m_I = +1, 0, -1)$. A spin $\frac{1}{2}$ nucleus (e.g., ^1H or ^{13}C) coupled to a single deuterium nucleus would therefore show a 1:1:1 triplet in its NMR spectrum. This is why the CDCl3 resonance is a 1:1:1 triplet in the ^{13}C NMR spectra of samples dissolved in $CDCl_3$.

In general, the number of lines that will be observed when spin $\frac{1}{2}$ nuclei couple to n nuclei with nuclear spin quantum number I is given by $n2I + 1$ (note that when $I = \frac{1}{2}$, this simplifies to the $n + 1$ rule we are already familiar with). When $n = 1$, the multiplet will always have lines of equal intensity (because the $2I + 1$ spin states are all equally likely). When $n > 1$, the intensity pattern follows a complicated binomial series.

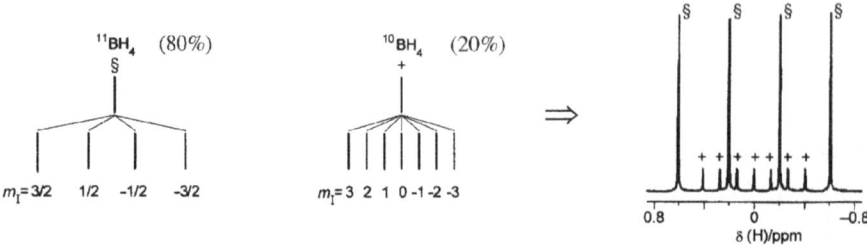

Example: The ^{1}H NMR spectrum of BH_4^- appears as a complex pattern of eleven lines. The appearance of the spectrum can be understood by considering that ^{10}B (I = 3) is 20% abundant and ^{11}B (I = $\frac{3}{2}$) is 80% abundant and applying the n2I+1 rule to each.

In theory, we could also record the ^{10}B or ^{11}B NMR spectra as well. In practice, ^{11}B NMR spectra are more frequently recorded because of its higher natural abundance and receptivity (i.e., a measure of how sensitive the nucleus is to having its NMR spectrum recorded). The 11B NMR spectrum would consist of a quintet in a 1:4:6:4:1 ratio as we would predict from the n+1 rule for coupling to four equivalent I = $\frac{1}{2}$ hydrogen nuclei. It is also interesting to note that $^1J_{11BH} \sim 3\left(^1J_{10BH}\right)$ since $\gamma\left(^{11}B\right) \sim 3\,\gamma\left(^{10}B\right)$.

(d) Linewidth and its effect on the appearance of NMR spectra: The linewidth is defined as the width of the peak at half its height, and is often measured in Hz. Sometimes, the linewidth of a resonance is larger than other NMR information, namely coupling, that may be of interest.

Example: An example of the influence on the linewidth of a quartet with J = 10 Hz is shown below. Note that when the linewidth is greater than J, the features of the multiplet are obscured.

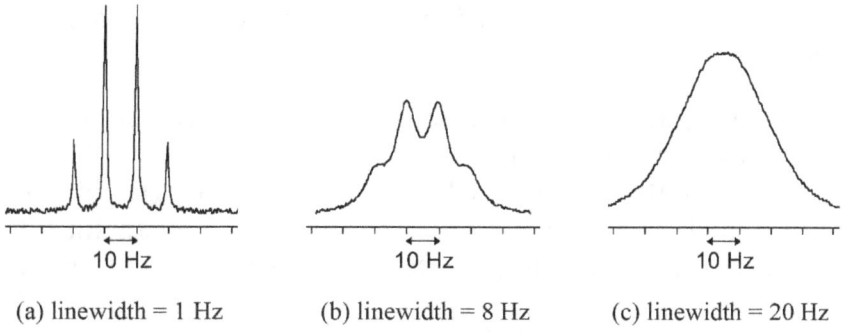

| (a) linewidth = 1 Hz | (b) linewidth = 8 Hz | (c) linewidth = 20 Hz |

(e) Second order spectra: Thus far we have discussed systems in which the magnitude of the coupling constants is much smaller than the difference in chemical shift between the coupled nuclei. When this criterion is in place, first-order spectra are observed that obey the n2I+1 rule and have intensity patterns that follow binomial series. First order spin systems are often labelled with letters that are widely spaced in the alphabet (e.g., AMX in Example 4) to denote the large chemical shift difference between the coupled nuclei.

However if coupled nuclei do not have greatly different chemical shifts (i.e., $\Delta v / J < \sim 10$ where Δv is the chemical shift difference in Hz, not ppm), second order spectra can result. In this case, the spin system is identified by letters that are closely spaced in the alphabet (e.g., AB or ABC), and the spectral line pattern that results from spin-spin coupling is not easily interpreted by inspection. Such spectra are best simulated using a program such as WinDNMR7 (available on the computer workstations in W1-50). The theory behind second order splitting patterns is beyond the scope of this course, and it is suggested that you experiment with the WinDNMR program if you wish to learn more about this phenomenon.

Example: The intensity pattern of an AB spin system changes quite dramatically as the chemical shift difference (Δv *in Hz*) decreases from 100 Hz to 20 Hz when JAB is held constant at 10 Hz. Note also that the chemical shift of A and B is no longer at the center of each "doublet". This AB system was simulated using WinDNMR.

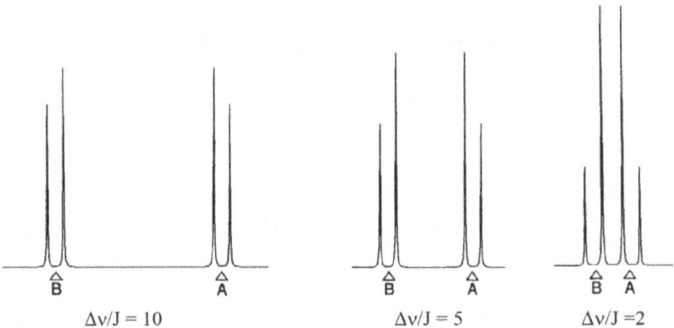

$\Delta v/J = 10$ $\Delta v/J = 5$ $\Delta v/J = 2$

(f) Magnetic versus chemical equivalence: It is possible that chemically equivalent nuclei may interact with adjacent nuclei such that the coupling constants have different magnitudes. When this occurs, the nuclei are said to be chemically equivalent but magnetically inequivalent.

Example: For example, in F_2POPF_2, the fluorine nuclei are chemically equivalent and the phosphorus nuclei are chemically equivalent; the ^{19}F NMR spectrum, however, is not a doublet.[8] Because the fluorine nuclei couple strongly to each of the phosphorus nuclei, both $^1J_{PF}$ and $^3J_{P'F}$ coupling are observed. In order to distinguish this fact, the spin system is labelled $X_2SS'X'_2$ to indicate the magnetic inequivalence of the chemically equivalent nuclei. When a situation like this occurs, the spectrum is invariably second order as $_{AA'}$, J_{AX}, $J_{A'X}$, and $J_{XX'}$ couplings can all contribute to the splitting pattern. In the case of F_2POPF_2, it was found that $^1J_{PF} = 1358$ Hz, $^3J_{P'F} = -14$ Hz,* $^2J_{PP'} = 4$ Hz, and $^5J_{FF'} \sim 0$ Hz.[8] It is important to note that if $^2J_{PP'}$ and $^3J_{P'F}$ were both negligibly small, the spectrum would reduce to a simple AX_2 pattern.

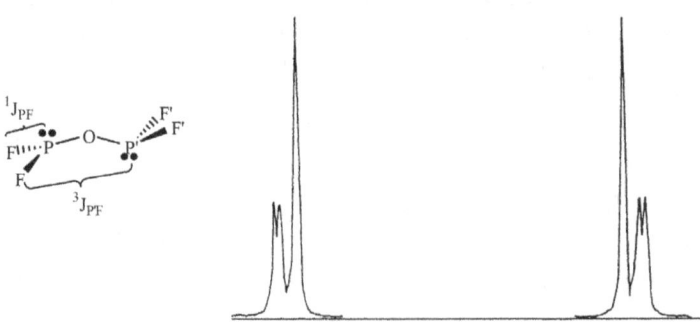

^{19}F NMR spectrum of F2POPF2.

Remember that chemically inequivalent nuclei are, by definition, magnetically inequivalent. Chemically equivalent nuclei are magnetically equivalent only if they are coupled to all other NMR active nuclei equally; this happens when the observed nucleus is related to each nucleus in any other set in the spin system by reflection through a mirror plane.

UV-Visible Spectroscopy

Recall that an absorption spectrum is obtained by recording the amount of radiation absorbed by the sample as a function of the frequency or wavelength of the incident radiation. When a molecule interacts with electromagnetic radiation, energy is absorbed and the molecule is promoted, or is said to undergo a transition, to a higher energy state (excited state). In order for absorption to occur, the energy of the radiation must match the energy difference between the quantized energy levels of the molecule.

The energy required for electronic transitions within molecules occurs in the UV and visible regions of the electromagnetic spectrum $(\lambda \sim 200 - 800\ nm)$. In transition metal compounds, absorptions in the visible and near-UV range frequently correspond to transitions of d electrons. The simplest case is a species with a single d electron in an octahedral geometry, such as $\left[Ti(OH_2)_6\right]^{3+}$. As is shown in figure below, absorption of radiation can promote the d electron from a t_{2g} orbital to an e_g orbital if the energy of the incident photon (E = hν) matches the energy spacing (Δ_o) between the t_{2g} and e_g orbitals. As a result, the UV-visible spectrum of $\left[Ti(OH_2)_6\right]^{3+}$ shows a single absorption with a maximum at roughly 510 nm. In this case, we could describe the transition as $t_{2g} \rightarrow e_g$.

incident radiation
(a photon where E=hν)

$$\Delta E = h\nu = \frac{hc}{\lambda}$$

h = Planck's constant, 6.623×10^{-34} J·s

c = speed of light, 3.00×10^{8} m·s^{-1}

UV-visible spectra are often reported in wavelength (λ) in units of nanometers (nm). Using the equation above and including Avagadro's number to convert to a molar quantity, the wavelength of the absorption maximum can be converted into an expression of energy absorbed by the sample. For example, given the absorption at 510 nm for $\left[Ti(OH_2)_6\right]^{3+}$, we would be able to estimate Δ_o :

$$\Delta E = h\nu = \frac{hc}{\lambda} = \frac{\left(6.623 \times 10^{-34}\ J.s\right)\left(3.00 \times 10^{8}\ m.s^{-1}\right)}{510 \times 10^{-9}} 6.022 \times 10^{23}\ mol^{-1} = 235\ kJ·mol^{-1}$$

The position of the UV-visible transition in figure also accounts for the red-violet colour of $\left[Ti(OH_2)_6\right]^{3+}$, because it is the transmitted light (the troughs in the UV-visible spectrum), not the absorbed light, that determines the colour of a compound. Furthermore, the colour of the transmitted light is generally the complement of the absorbed colour. A simple way of relating the absorption wavelength to the colour of the compound is to use an artist's colour wheel.

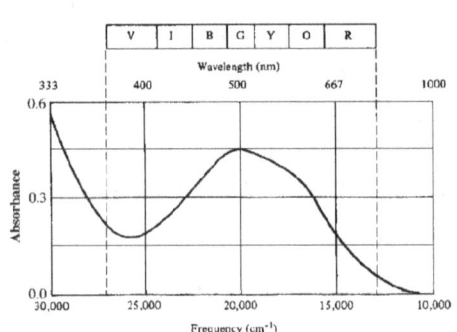

Figure: UV-visible spectrum of 0.1 M
$[Ti(OH_2)_6]^{3+}_{(aq)}$

Artist's color wheel

From the standpoint of crystal field theory, the position of the ligands in the spectrochemical series will affect the magnitude of the crystal field splitting (e.g., Δ_o for octahedral complexes). Changes in Δ_o affect the energy at which d-d electronic transitions occur, which consequently alters the colour of the complex. The spectrochemical series for selected ligands is shown below:

$$I^- < Br^- < \underline{S}CN < Cl^- < NO_3^- < F^- < OH^- < H_2O < \underline{N}CS^- < NO_2^- < PR_3 < CN^- < CO$$

weak field, small Δ_o strong field, large Δ_o

The intensity of the absorption determines whether the colour will be pale or dark. Beer's Law relates the absorbance of a peak to the molar absorptivity or extinction coefficient, ε, of each band in the UV-visible spectrum (i.e., each absorbance maximum, λ_{max}, has an associated value of ε):

$A = \varepsilon \ell c$ where A = absorbance (vertical scale of the spectrum; should ideally
                     not exceed a value of 1 for a "good" spectrum)
            e = the extinction coefficient $\left(\text{in cm}^{-1}\cdot M^{-1}\right)$
            l = the path length of the cell $(= 1.0$ cm in many instruments$)$
            c = the concentration of the solution (in M; should ideally be
                     chosen to provide an absorbance of between $0.6 - 0.8$ for the main peak(s) of interest)

As with other types of scientific data, the information extracted from a UV-visible spectrum can be conveniently summarized in a table which should accompany the labeled (name, date, compound, solvent, concentration) spectrum you tape into your notebook (e.g., for the spectrum in figure recorded in a 1.0 cm path length cell):

Compound*	λmax (nm)	$\varepsilon\,(cm^{-1}\cdot M^{-1})$
$[Ti(OH2)6]^{3+}$ (aq)	510	4.5

The magnitude of the extinction coefficient (intensity of the colour) is affected by two quantum mechanical selection rules that state whether transitions are allowed (intense colour) or forbidden (pale colour).

1) Spin selection rule –the number of unpaired electrons in a molecule cannot change upon excitation (i.e., an electron "spin-flip" is forbidden).

2) Symmetry selection rule - if the molecule has a center of symmetry (inversion center), transitions from one centrosymmetric orbital to another are forbidden (i.e., all d orbitals are centrosymmetric, so d-d transitions are forbidden in a complex that has an inversion center. If the molecule does not have an inversion center, d-d transitions are symmetry allowed).

For example, octahedral $\left[Ti(OH_2)_6\right]^{3+}$ has a center of symmetry and thus has a low extinction coefficient $\left(\sim 4.5\,cm^{-1}\cdot M^{-1}\right)$ because the transition is symmetry forbidden. However, no spin flip results, so the transition is spin allowed. Symmetry and spin allowed transitions typically have extinction coefficients greater than 1000 $cm^{-1}\cdot M^{-1}$. As a result of the symmetry selection rule, octahedral complexes generally have extinction coefficients between 1 and 100 $cm^{-1}\cdot M^{-1}$.

Systems with more than one d electron present additional complexities. For example, in high spin d^5 $\left[Mn(OH_2)_6\right]^{2+}$, each of the t_{2g} and e_g orbitals is occupied by one electron. Accordingly, d-d transitions are both spin and symmetry forbidden, and solutions of Mn(II) appear as a very pale pink colour with extinction coefficients less than 0.1 $cm^{-1}\cdot M^{-1}$. The example of $\left[Mn(OH_2)_6\right]^{2+}$ is still somewhat simplistic, however, for a multi-d electron system.

In a d^1 octahedral complex, the d orbitals are split into two sets, t_{2g} and e_g, as a result of the interaction of the ligands with the metal-based orbitals, and the single d electron can reside in either of these sets (i.e., in a t_{2g} orbital in the ground state or an e_g orbital in the excited state). If we were to add another electron into either of these orbitals, we would now have to worry about interactions between the electrons in addition to the orbital splitting caused by the octahedral ligand field. The interactions between the electrons produce a number of possible electronic states, which can loosely be defined as the different ways that the electrons can distribute themselves among the available orbitals. Suffice it to say that the number of possible electronic transitions are not easily predicted by referring to an orbital splitting diagram; for instance, an octahedral d^2 complex has three spin-allowed transitions.[1] In systems where there are a number of electronic states available, it is not correct to describe the transitions as occurring from one orbital to another (rather, the transition is from one electronic state to another). The concept of electronic states and their influence on the appearance of electronic spectra is covered in more detail in Reference[1].

Despite the added complexity of electronic states, we can still make use of UV-visible spectra to assign coordination geometries by drawing on past precedent in the literature. For example, $\left[Co(OH_2)_6 \right]^{3+}$ is a d^6 system which is low spin with three possible electronic states, designated as $^1A_{1g}$, $^1T_{1g}$, and $^1T_{2g}$. Consequently, two transitions are observed in its UV-visible spectrum (Figure 12(b) where X = H_2O). Now, if we replace two of the "X" ligands with a different ligand "Y", two possible geometric isomers result, cis-$\left[CoX_4Y_2 \right]$ and trans-$\left[CoX_4Y_2 \right]$. It turns out that the $^1T_{1g}$ electronic state splits when this occurs, providing two new electronic states. As a result, three UV-visible transitions are theoretically expected. However, the splitting is much smaller for the case where the "Y" ligands are cis, and thus two of the bands $(v_1 \sim v_2)$ voccur at very similar frequencies and overlap. In the case where the "Y" ligands are trans, the splitting is much greater, providing three distinct bands in the UV-visible spectrum. It is important to remember that the features discussed above apply specifically to low spin d6 systems based on an octahedral geometry; for example, a d4 system would behave differently, as would d6 systems with other coordination geometries.

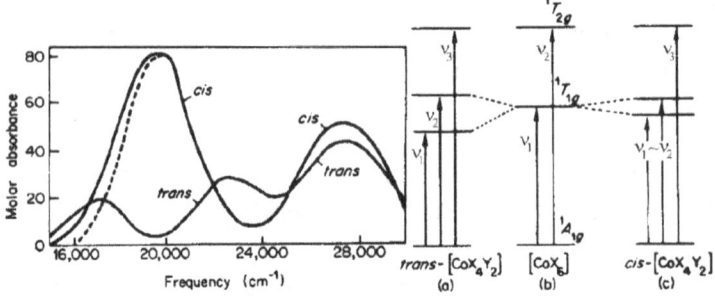

An example of a multi-d electron system for which UV-visible transitions can be more easily described is square planar d8 (e.g., Ni^{2+}) ML_4 complexes. As shown in figure, three spin-allowed transitions are theoretically possible. In practice, v1 is often the only band that can be observed in the visible region; this is quite useful from the point of view of crystal field theory since $v_1 = \Delta$ in this system. The v_2 and v_3 bands are higher in energy (shorter λ) and are frequently masked by other high-energy electronic transitions.

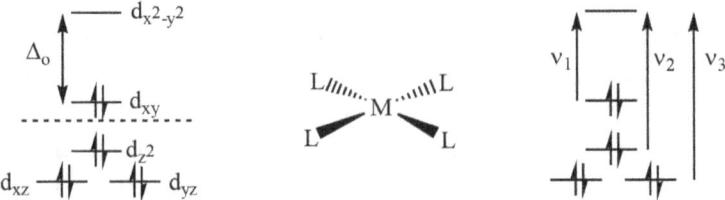

Spin-allowed electronic transitions in a square planar d^8 complex

While it is true that d-d transitions are usually the most prominent and important features of the electronic spectra of transition metal complexes, another class of electronic transitions (called charge transfer or CT transitions) are frequently observed in the high energy (low wavelength) region of the UV-visible spectrum.1 Recall that d-d transitions involve a redistribution of electrons within the d orbitals. CT transitions generally result when an electron is transferred between metal and ligand based orbitals, and usually require more energy than d-d transitions. Unlike d-d transitions, those involving charge transfer are fully allowed. As a result, CT bands often have molar absorptivities about 1000 times greater than those for d-d transitions. When these absorptions fall within the visible range of the spectrum, they often produce rich colours.1 Thus, colours in transition metal compounds are not always associated with d-d transitions. For example, the permanganate ion, MnO_4^- (Mn^{+7}, d^0) is a deep purple colour as a result of charge transfer between filled ligand orbitals and empty d-orbitals on the metal ion.

X-ray Crystallography

Single-crystal X-ray diffraction is the most powerful X-ray technique for inorganic chemists. From precise measurement of the intensity and angles at which an X-ray beam diffracts off a crystal, the arrangement of the atoms can be reconstructed. Obviously, as a direct probe of structure crystallography is an invaluable characterization method for all types of compounds. Some inorganic compounds (e. g., rocks, minerals) can't be obtained as single crystals. In these cases X-ray powder diffraction can be used to obtain the dimensions of the unit cell for use in identification (there is a large, indexed catalog of lattice constants for many minerals).

X-ray Crystallography is a scientific method used to determine the arrangement of atoms of a crystalline solid in three dimensional space. This technique takes advantage of the interatomic spacing of most crystalline solids by employing them as a diffraction gradient for x-ray light, which has wavelengths on the order of 1 angstrom $(10^{-8}\, cm)$.

In 1895, Wilhelm Rontgen discovered x- rays. The nature of x- rays, whether they were particles or electromagnetic radiation, was a topic of debate until 1912. If the wave idea was correct, researchers knew that the wavelength of this light would need to be on the order of 1 Angstrom (A) (10^{-8} cm). Diffraction and measurement of such small wavelengths would require a gradient with spacing on the same order of magnitude as the light.

In 1912, Max von Laue, at the University of Munich in Germany, postulated that atoms in a crystal lattice had a regular, periodic structure with interatomic distances on the order of 1 A. Without having any evidence to support his claim on the periodic arrangements of atoms in a lattice, he further postulated that the crystalline structure can be used to diffract x-rays, much like a gradient in an infrared spectrometer can diffract infrared light. His postulate was based on the following assumptions: the atomic lattice of a crystal is periodic, x- rays are electromagnetic radiation, and the interatomic distance of a crystal are on the same order of magnitude as x- ray light. Laue's predictions were confirmed when two researchers: Friedrich and Knipping, successfully photographed the diffraction pattern associated with the x-ray radiation of crystalline $CuSO_4 \cdot 5H_2O$. The science of x-ray crystallography was born.

The arrangement of the atoms needs to be in an ordered, periodic structure in order for them to diffract the x-ray beams. A series of mathematical calculations is then used to produce a diffraction pattern that is characteristic to the particular arrangement of atoms in that crystal. X-ray crystallography remains to this day the primary tool used by researchers in characterizing the structure and bonding of organometallic compounds.

Diffraction

Diffraction is a phenomena that occurs when light encounters an obstacle. The waves of light can either bend around the obstacle, or in the case of a slit, can travel through the slits. The resulting diffraction pattern will show areas of constructive interference, where two waves interact in phase, and destructive interference, where two waves interact out of phase. Calculation of the phase difference can be explained by examining figure below.

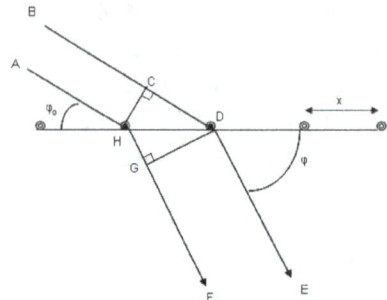

Scattering of light by a diffraction gratting

In the figure below, two parallel waves, BD and AH are striking a gradient at an angle θ_o. The incident wave BD travels farther than AH by a distance of CD before reaching the gradient. The scattered wave (depicted below the gradient) HF, travels father than the scattered wave DE by a distance of HG. So the total path difference between path AHGF and BCDE is CD - HG. To observe a wave of high intensity (one created through constructive interference), the difference CD - HG must equal to an integer number of wavelengths to be observed at the angle psi, $CD - HG = n\lambda$, where λ is the wavelength of the light. Applying some basic trigonometric properties, the following two equations can be shown about the lines:

$$CD = xcos(\theta o)$$

And

$$HG = x cos(\theta)$$

where x is the distance between the points where the diffraction repeats. Combining the two equations,

$$x(cos\theta_o - cos\theta) = n\lambda$$

Bragg's Law

Diffraction of an x-ray beam, occurs when the light interacts with the electron cloud surrounding the atoms of the crystalline solid. Due to the periodic crystalline structure of a solid, it is possible to describe it as a series of planes with an equal interplaner distance. As an x-ray's beam hits the surface of the crystal at an angle ?, some of the light will be diffracted at that same angle away from the solid. The remainder of the light will travel into the crystal and some of that light will interact with the second plane of atoms. Some of the light will be diffracted at an angle theta, and the remainder will travel deeper into the solid. This process will repeat for the many planes in the crystal. The x-ray beams travel different pathlengths before hitting the various planes of the crystal, so after diffraction, the beams will interact constructively only if the path length difference is equal to an integer number of wavelengths (just like in the normal diffraction case above). In the figure below, the difference in path lengths of the beam striking the first plane and the beam striking the second plane is equal to BG + GF. So, the two diffracted beams will constructively interfere (be in phase) only if $BG + GF = n\lambda$. Basic trigonometry will tell us that the two segments are equal to one another with the interplaner distance times the sine of the angle θ. So we get:

$$BG = BC = d\ sin\ \theta$$

Thus,

$$2d\ sin\ \theta = n\lambda$$

This equation is known as Bragg's Law, named after W. H. Bragg and his son, W. L. Bragg; who discovered this geometric relationship in 1912. {C}{C}Bragg's Law relates the distance between two planes in a crystal and the angle of reflection to the x-ray wavelength. The x-rays that are diffracted off the crystal have to be in-phase in order to signal. Only certain angles that satisfy the following condition will register:

$$sin\ \theta = \frac{n\lambda}{2d}$$

For historical reasons, the resulting diffraction spectrum is represented as intensity vs. 2θ.

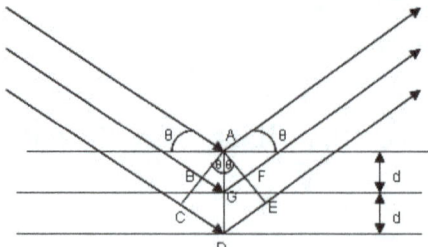

Incident X-ray strikes a set of planes (with an interplaner distance of d) at an angle of θ

Instrument Components

The main components of an x-ray instrument are similar to those of many optical spectroscopic instruments. These include a source, a device to select and restrict the wavelengths used for measurement, a holder for the sample, a detector, and a signal converter and readout. However, for x-ray diffraction; only a source, sample holder, and signal converter/readout are required.

The Source

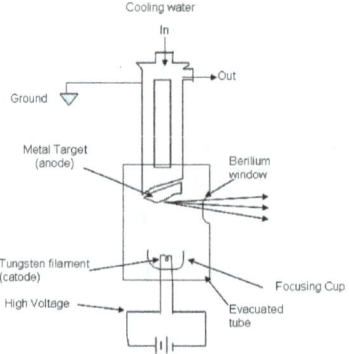

x-ray tubes schematic

x-ray tubes provides a means for generating x-ray radiation in most analytical instruments. An evacuated tube houses a tungsten filament which acts as a cathode opposite to a much larger, water cooled anode made of copper with a metal plate on it. The metal plate can be made of any of the following metals: chromium, tungsten, copper, rhodium, silver, cobalt, and iron. A high voltage is passed through the filament and high energy electrons are produced. The machine needs some way of controlling the intensity and wavelength of the resulting light. The intensity of the light can be controlled by adjusting the amount of current passing through the filament; essentially acting as a temperature control. The wavelength of the light is controlled by setting the proper accelerating voltage of the electrons. The voltage placed across the system will determine the energy of the electrons traveling towards the anode. X-rays are produced when the electrons hit the target metal. Because the energy of light is inversely proportional to wavelength $(E = hc = h(1/\lambda))$, controlling the energy, controls the wavelength of the x-ray beam.

X-ray Filter

Monochromators and filters are used to produce monochromatic x-ray light. This narrow wavelength range is essential for diffraction calculations. For instance, a zirconium filter can be used to cut out unwanted wavelengths from a molybdenum metal target. The molybdenum target will produce x-rays with two wavelengths. A zirconium filter can be used to absorb the unwanted emission with wavelength $K\beta$, while allowing the desired wavelength, $K\alpha$ to pass through.

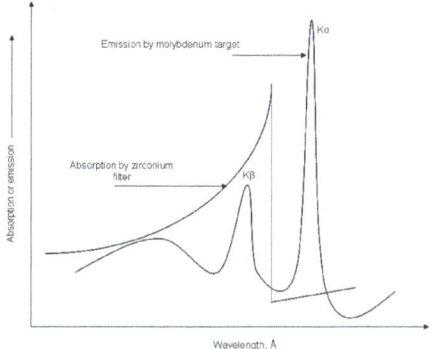

Figure 4. Monochromatic light produce by absorption of unwanted emissions

Needle Sample Holder

The sample holder for an x-ray diffraction unit is simply a needle that holds the crystal in place while the x-ray diffractometer takes readings.

Signal Converter

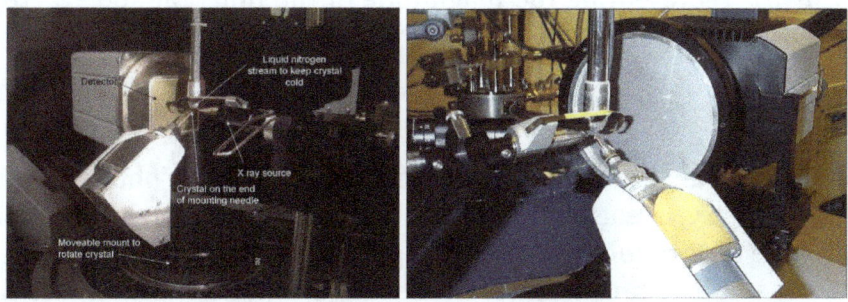

Figure: (left) A view of the entire machine and (right) a crystal mounted on a goniometer shown with the x-ray generator and detector

In x-ray diffraction, the detector is a transducer that counts the number of photons that collide into it. This photon counter gives a digital readout in number of photons per unit time. Below is a figure of a typical x-ray diffraction unit with all of the parts labeled.

Fourier Transform

In mathematics, a Fourier transform is an operation that converts one real function into another. In the case of FTIR, a Fourier transform is applied to a function in the time domain to convert it into the frequency domain. One way of thinking about this is to draw the example of music by writing it down on a sheet of paper. Each note is in a so-called "sheet" domain. These same notes can also be expressed by playing them. The process of playing the notes can be thought of as converting the notes from the "sheet" domain into the "sound" domain. Each note played represents exactly what is on the paper just in a different way. This is precisely what the Fourier transform process is doing to the collected data of an x-ray diffraction. This is done in order to determine the electron density around the crystalline atoms in real space. The following equations can be used to determine the electrons' position:

$$p(x,y,z) = \sum_{h}\sum_{k}\sum_{l} lF(hkl)e^{-2\pi i(hx+ky+lz)}$$

$$\int_{0}^{1}\int_{0}^{1}\int_{0}^{1} p(x,y,z)e^{2\pi i(hx+ky+lz)}dx\,dy\,dz$$

$$F(q) = |F(q)|e^{i\phi(q)}$$

where $p(xyz)$ is the electron density function, and $F(hkl)$ is the electron density function in real space. Equation ($p(x,y,z) = \sum_{h}\sum_{k}\sum_{l} lF(hkl)e^{-2\pi i(hx+ky+lz)}$) represents the Fourier expansion of the electron density function. To solve for $F(hkl)$, the equation $p(x,y,z) = \sum_{h}\sum_{k}\sum_{l} lF(hkl)e^{-2\pi i(hx+ky+lz)}$ needs to be evaluated over all values of h, k, and l, resulting in Equation ($\int_{0}^{1}\int_{0}^{1}\int_{0}^{1} p(x,y,z)e^{2\pi i(hx+ky+lz)}dx\,dy\,dz$). The resulting function $F(hkl)$ is generally expressed as a complex number (as seen in equation $F(q) = |F(q)|e^{i\phi(q)}$ above) with $|F(q)|$ representing the magnitude of the function and ϕ representing the phase.

Crystallization

In order to run an x-ray diffraction experiment, one must first obtain a crystal. In organometallic chemistry, a reaction might work but when no crystals form, it is impossible to characterize the products. Crystals are grown by slowly cooling a supersaturated solution. Such a solution can be made by heating a solution to decrease the amount of solvent present and to increase the solubility of the desired compound in the solvent. Once made, the solution must be cooled gradually. Rapid temperature change will cause the compound to crash out of solution, trapping solvent and impurities within the newly formed matrix. Cooling continues as a seed crystal forms. This crystal is a point where solute can deposit out of the solution and into the solid phase. Solutions are generally placed into a freezer (-78 °C) in order to ensure all of the compound has crystallized. One way to ensure gradual cooling in a -78 °C freezer is to place the container housing the compound into a beaker of ethanol. The ethanol will act as a temperature buffer, ensuring a slow decrease in the temperature gradient between the flask and the freezer. Once crystals are grown, it is imperative that they remain cold as any addition of energy will cause a disruption of the crystal lattice, which will yield bad diffraction data. The result of an organometallic chromium compound crystallization can be seen below.

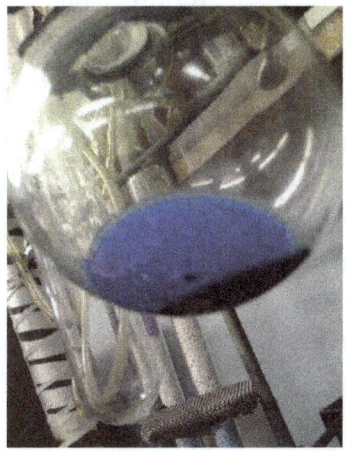

Figure: Organometallic chromium crystals in a Schlenk under nitrogen

Mounting the Crystal

Due to the air-sensitivity of most organometallic compounds, crystals must be transported in a highly viscous organic compound called paratone oil (see picture below). Crystals are abstracted from their respective Schlenks by dabbing the end of a spatula with the paratone oil and then sticking the crystal onto the oil. Although there might be some exposure of the compounds to air and water, crystals can withstand more exposure than solution (of the preserved protein) before degrading. On top of serving to protect the crystal, the paratone oil also serves as the glue to bind the crystal to the needle.

Rotating Crystal Method

To describe the periodic, three dimensional nature of crystals, the Laue equations are employed:

$$a(cos\theta_o - cos\theta) = h\lambda$$

$$b(cos\theta_o - cos\theta) = k\lambda$$

$$c(cos\theta_o - cos\theta) = l\lambda$$

where a, b, and c are the three axes of the unit cell, θ_o ,o , ?o are the angles of incident radiation, and ?, ?, ? are the angles of the diffracted radiation. A diffraction signal (constructive interference) will arise when h, k, and l are integer values. The rotating crystal method employs these equations. X-ray radiation is shown onto a crystal as it rotates around one of its unit cell axis. The beam strikes the crystal at a 90 degree angle. Using equation 1 above, we see that if θ_o is 90 degrees, then $cos\ \theta_o = 0$. For the equation to hold true, we can set h=0, granted that\theta= 90. The above three equations will be satisfied at various points as the crystal rotates. This gives rise to a diffraction pattern (shown in the image below as multiple h values). The cylindrical film is then unwrapped and developed. The following equation can be used to determine the length axis around which the crystal was rotated:

$$a = \frac{ch\lambda}{\sin\ tan^{-1}(y/r)}$$

where a is the length of the axis, y is the distance from $h=0$ to the h of interest, r is the radius of the firm, and ? is the wavelength of the x-ray radiation used. The first length can be determined with ease, but the other two require far more work, including re-mounting the crystal so that it rotates around that particular axis.

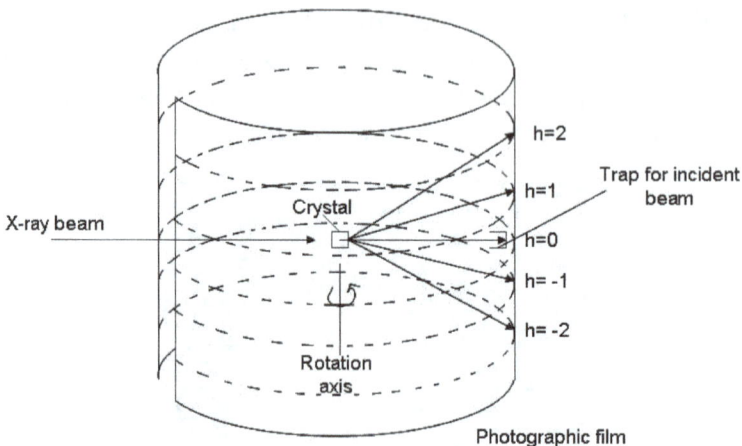

X-ray Crystallography of Proteins

The crystals that form are frozen in liquid nitrogen and taken to the synchrotron which is a highly powered tunable x-ray source. They are mounted on a goniometer and hit with a beam of x-rays. Data is collected as the crystal is rotated through a series of angles. The angle depends on the symmetry of the crystal.

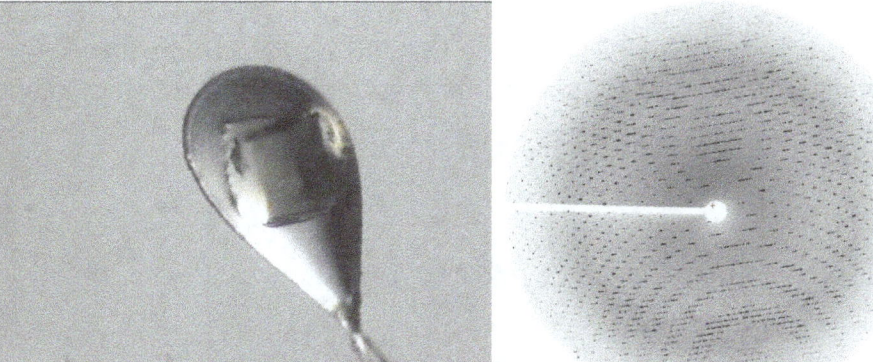

Figure: This is a picture of a protein crystal mounted on a loop with respect to the UC Davis Structural Biology Lab; Bottom Right) This is a diffraction pattern created from the APS Kinase D63N Mutant of the above crystal with respect to the UC Davis Structural Biology Lab

Proteins are among the many biological molecules that are used for x-ray Crystallography studies. They are involved in many pathways in biology, often catalyzing reactions by increasing the reaction rate. Most scientists use x-ray Crystallography to solve the structures of protein and to determine functions of residues, interactions with substrates, and interactions with other proteins or nucleic acids. Proteins can be co - crystallized with these substrates, or they may be soaked into the crystal after crystallization.

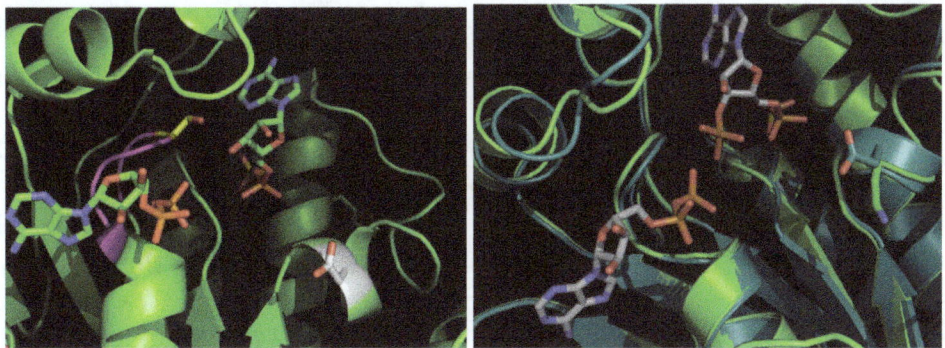

Figure: This is the structure of APS Kinase co - crystallized with ligands ADP and APS created via pymol by an undergrad working in the Structural Biology lab at UC Davis; Bottom right) This is the mutant overlay of APS kinase.

The teal is the wild - type and the lime green is the mutant. D63 (from the wild-type) is mutated to asparagine. Images created by pymol by an undergrad working in the Structural Biology lab at UC Davis.

Protein crystallization

Proteins will solidify into crystals under certain conditions. These conditions are usually made up of salts, buffers, and precipitating agents. This is often the hardest step in x-ray crystallography. Hundreds of conditions varying the salts, pH, buffer, and precipitating agents are combined with the protein in order to crystallize the protein under the right conditions. This is done using 96 well plates; each well containing a different condition and crystals; which form over the course of days, weeks, or even months. The pictures below are crystals of APS Kinase D63N from *Penicillium chrysogenum* taken at the Chemistry building at UC Davis after crystals formed over a period of a week.

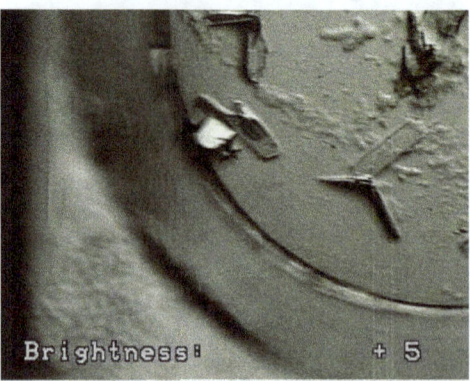

Figure: The experiment was originally proposed by Dr. Andy Fisher in the Structural Biology lab. Michelle Towles was a research assistant to Sean Gay and purified the protein and set up the crystal trays.

Dual Polarization Interferometer

Dual Polarization Interferometry (DPI) DPI is currently one of the most powerful label-free biosensing techniques to record real-time data of conformational dynamics. That fact has been efficiently employed in different applications, such as bionanotechnology, crystallography, surface science and drug discovery. Data provided by this technique enable the understanding of the intimate links between the structural changes in biomolecules and their functions and interactions. Consequently this technique can be considered as one of the most well-built and well-thought through waveguide sensors. Hence, the use of an interferometric readout and a long waveguide makes the measurements very stable and more accurate than those of the competitor techniques. Despite this, it has its weaknesses, e.g. the necessity of using a relatively long sensing element, which makes SPR more accurate when studying initial kinetics of adsorption to surfaces.

General Principles

In a DPI instrument the laser light broadly illuminates the end face of the two stacked planar waveguides. A small portion of the light is coupled into each of the two waveguides; one acting as a sensing surface, while the second is used as a reference beam. As a result, on emerging from the waveguide structure, a two-dimensional interference pattern is formed in the far field by combining the light passing through both waveguides. Moreover, the waveguide stack is designed to allow both measurement and reference arm to support single modes in two different polarizations -transverse-electric (TE) and transverse-magnetic (TM)-, enabling the measurement of two optical phase changes. The polarization can be switched rapidly (on a 2 ms cycle), allowing real-time measurements of molecular processes taking place on a chip surface in a flow-through system.

The input light polarization state is switched between TE and TM excitation modes by a liquid crystal wave plate at 500 Hz. The output far-field Young's fringe pattern formed from the slab waveguide modes provides, by Fourier transformation, the accumulated real-time mode fringe phase changes, $\Delta\Phi TE$ and $\Delta\Phi TM$, during the deposition of the layer. Typically, these phase changes are used to calculate solutions to Maxwell's equations for a one-dimensional multilayer dielectric continuum electrodynamic model, which finds the effective medium refractive index and thickness of the surface bound equivalent layer. The described process is used to obtain layer thickness and refractive index measurements throughout the experiment. The value of refractive index can be transformed into density of the material bound to the surface, provided its nature is known (proteins, nucleic acids, lipid membranes), and combining density with thickness, the surface mass can also be determined.

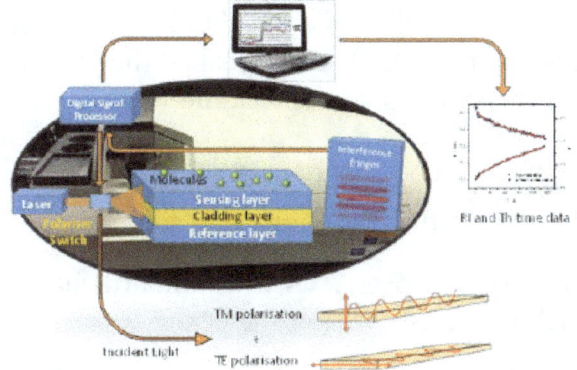

Figure: Schematic representation of a typical DPI sensor.

Furthermore, DPI allows the adlayer anisotropy, the quality of exhibiting properties with different values when measured along axes in different directions, to be measured in real-time. So, the material nonuniformity can be characterized by birefringence, Δn, which is the optical property of a material having a refractive index that depends on the polarization and propagation direction of light.

Applications

One of the most immediate applications of DPI was the protein characterization, which is a very important issue that should be addressed in the development of new pharmaceuticals and biomarkers. For instance, in an early publication DPI and AFM were employed to measure the dimensions of C-reactive protein. But new applicabilities of this technique in the protein world appeared soon, such as control of protein crystallization in a tandem DPI-dialysis, and the monitoring of protein adsorption, e.g. lysozyme, on the chip surface; in that study DPI was used in combination with QCM-D in order to calculate not only the protein adsorbed but also the adhered solvent mass. However, bearing in mind the increasing use of antibodies for biosensing, probably the most important application is the study of the antigen-antibody affinity binding, and its conformational dynamics.

Another important field of application is the monitoring of nucleic acid interactions, not only hybridization but also immobilization and non-specific binding. The study of phenomena taking place on lipid membranes is also a very popular and useful application, due to the need to develop in vitro environments that mimic cell membranes; it has allowed important applications such as the monitoring of the membrane disruption caused by peptides having antibiotic properties, performed through birefringence measurments.

Electrochemistry

Electrochemistry is the study of chemical processes that cause electrons to move. This movement of electrons is called electricity, which can be generated by movements of electrons from one element to another in a reaction known as an oxidation-reduction ("redox") reaction. Cyclic voltammetry has been called the "electrochemical equivalent of spectroscopy" because it can be used to determine

1) The oxidation state,

2) Eo of each redox process that the compound can undergo and

3) The kinetics of the redox process. CV is a standard characterization method in most inorganic laboratories.

There are two types of electrochemical cells: galvanic, also called Voltaic, and electrolytic. Galvanic cells derives its energy from spontaneous redox reactions, while electrolytic cells involve non-spontaneous reactions and thus require an external electron source like a DC battery or an AC power source. Both galvanic and electrolytic cells will consist of two electrodes (an anode and a cathode), which can be made of the same or different metals, and an electrolyte in which the two electrodes are immersed.

Galvanic Cells

Galvanic cells traditionally are used as sources of DC electrical power. A simple galvanic cell may contain only one electrolyte separated by a semi-porous membrane, while a more complex version involves two separate half-cells connected by a salt bridge. The salt bridge contains an inert electrolyte like potassium sulfate whose ions will diffuse into the separate half-cells to balance the building charges at the electrodes. According to the mnemonic "Red Cat An Ox", oxidation occurs at the anode and reduction occurs at the cathode. Since the reaction at the anode is the source of electrons for the current, the anode is the negative terminal for the galvanic cell.

Let's look at an example of a galvanic cell like the classic AA alkaline battery, in which the equations for the two half-reactions are listed below:

$$MnO_{2(s)} + H_2O + e^- \rightarrow MnOOH_{(s)} + OH^-_{(aq)} \; ; E^o = +0.382V$$

$$Zn(OH)_{2(s)} + 2e^- \rightarrow Zn_{(s)} + 2OH^-_{(aq)} \; ; E^0 = -1.221V$$

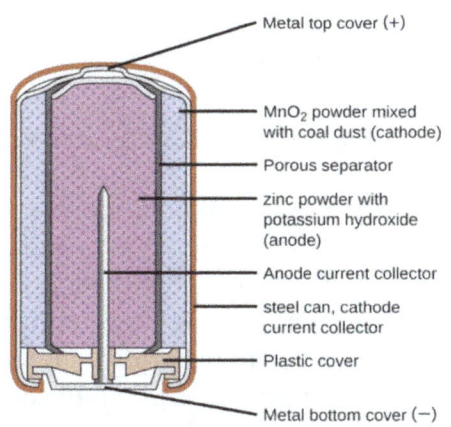

Diagram of AA alkaline battery cross section

Here we are given two reduction potentials for the anode and cathode. Most data tables listing cell potentials will list reduction potentials. Since they are both reduction potentials, we need to decide which one needs to be flipped. Let's keep in mind that when we flip the reduction potential into an oxidation potential, we also need to flip the sign. Since galvanic cells have a positive EMF, we are looking to flip the equation that when added to the other EMF will give us a positive value. By flipping the zinc half-reaction, we have the two EMF values as +0.382 V and +1.221 V. To find the cell EMF, we simply add them together to give us an approximate value of 1.5 V, which we already know to be the EMF of an alkaline AA battery.

Since voltage is an intensive property, which is one that does not depend on the system size or the amount of material in the system, we do not have to multiply the EMF by any stoichiometric coefficient to cancel out the electrons in calculating EMF.

Calculating the Gibbs Free Energy from EMF

Suppose we are asked to put the amount of energy into more thermodynamic terms. Let's use the following equation to connect E^o to ΔG^o :

$$\Delta G^o = -nFE^o$$

where n is the moles of electrons transferred, E^o is the standard state EMF, and F is Faraday's constant, 96,485 C/mol e^- but we will approximate as 10^5, C/mol e^- One way to remember this equation is to compare it to the formula for electric potential energy, U = qV. U can be correlated with ΔG^o start superscript, o, end superscript, both expressed in joules, and E^o with V, both expressed in volts. nF cancels out to be coulombs, just as q is.

We can start by plugging in values for an alkaline battery into our equation for the standard Gibbs free energy:

$$\Delta G^o = -nFE^o$$

$$\Delta G^o = -(2\,mol\,e^-)\left(10^5\,\frac{C}{mol\,e^-}\right)(1.5V)$$

$$\Delta G^o = -300,000\,J \ \ or -300\,kJ$$

Gibbs free energy is normally expressed in kilojoules rather than joules. The sign gives us an indication of the direction that the reaction has to shift to reach equilibrium. This means that a system under standard conditions would have to shift to the right, converting some reactants into products before reaching equilibrium. The magnitude gives us an indication of how far the standard state is from equilibrium. The larger the value, the further the reaction has to shift from standard state conditions to reach equilibrium. A negative value for ΔG^o also indicates the spontaneity of the reaction.

Calculating the Equilibrium Constant from EMF

Suppose we are asked to calculate the equilibrium constant K at standard conditions to get an indication of how favorable this reaction is.

$$\Delta G^o = -RT \ ln \ K_{eq}$$

$$\Delta G^o = -\left(8.31\frac{J}{mol * K}\right)(298\,K)(2.303)\log K$$

Let's approximate the gas constant R with the value of 8.31 J/(mol·K) as 8 and the standard temperature of 298 K as 300, which when multiplied together calculates to a value of 2400. Since we rounded down the original value of R earlier, the actual calculated value should be closer to 2500, which we will use instead. We need to multiply R by the conversion factor for lnx to logx, which is 2.303. We would not be expected to know this value and would most likely be provided the conversion factor, which we will round down to 2.3. In order to calculate (2500 x 2.3), use the values of 2500 x 2 and 2500 x 3 as the lower and upper boundaries, respectively 5000 and 7500. 2.3 is closer to 2 than 3, so let's find a value closer to 5000 than 7500, which would be 6000.

$$\Delta G^o \approx -(8)(300)(2.3)\log K \approx -2500(2.3)log \ K \approx -6000 * \log K$$

$$-300,000 = -6000 * \log K$$

$$\frac{300000}{5800} \approx \frac{300}{6} = 50 = \log \ K$$

$$10^{\log K} = K = 10^{50}$$

We can plug in the value of ΔG^{o} on the left side of the equation. Even though ΔG^{o} is normally expressed as kJ/mol, R is expressed as J/mol·K, so we can convert R or ΔG^{o} to match units. Let's plug in 300,000 J for ΔG^{o} to match R. Divide 300,000 by 6,000 to obtain a value of 50. To get rid of the logarithm, we raise both sides as exponents with a base of 10. Our calculated equilibrium constant is approximately 10^{50}, while the actual value is 10^{52}. In matching our answer to an answer choice, we are not looking an exact match. For example, answer choices that we may encounter are the following: $10^{52}, 10^{-52}, 10^{42}$, and 10^{62}.

The large K value indicates that the reaction is very favorable towards the products and will go entirely to completion. In the case of the batteries, the reaction will run until it reaches equilibrium, i.e. $\Delta G^{o} = 0$.

Concentration Cells and Nernst Equation

Let's consider a concentration cell, which is a specific form of a galvanic cell with two equivalent half-cells of the same material differing only in concentration. We can find concentration cells in the concentration gradients in our nerve cells, the Na^{+}/k^{+} or Ca^{2+} ion pumps in our cell membranes, and ATP synthase used in energy production.

A concentration cell produces a small voltage as it attempts to reach equilibrium. We can calculate the potential developed by a concentration cell using the Nernst Equation, which is as follows:

$$E = E^{o} - \frac{RT}{nF} lnQ, \text{ where } Q = \frac{[products]^{b}}{[reactants]^{a}} = \frac{[C]^{c}[D]^{d}}{[A]^{a}[B]^{b}}$$

where E is cell potential of interest, E^{o} is the standard cell potential, R is the gas constant, T is the absolute temperature, n is the moles of electrons transferred in the reaction, F is Faraday's constant, and Q is the reaction quotient.

We can take note how the Nernst equation fits the rubric for any equation with such a wide-ranging set of values that requires that it be expressed in a logarithmic scale: calculated value = standard or reference value ± logarithmic term. Let's compare this to other equations with a logarithmic term, like the Henderson-Hasselbalch equation and, of course, the thermodynamics equation from which the Nernst equation is derived:

$$pH = pK_{a} + \log \frac{[A]}{[HA]}$$
$$\Delta G = \Delta G^{o} + RT \ln Q$$

We can take notice how the equation matches the algebraic formula for a linear graph: y = mx + b, where y is E, m is -RT/nF, x is lnQ, and b is E^{o}.

For concentration cells, the Nernst equation should be simplified to the following:

$$E = E^\circ - \frac{0.06}{n}\left(\log\frac{[ion]_{low}}{[ion]_{high}}\right)$$

since (RT/F)(ln conversion factor) is equivalent to 0.0592 V, which we will round up to 0.06 V. As for the logarithmic term, [ion]/[ion] represents the reaction quotient Q. When Q = 1, the cell reaches equilibrium and is considered "dead" since the two concentrations are equal. In addition, log(1) = 0 so that E = 0, which confirms that the cell would be "dead". When $[ion]_{low}$ is in the numerator, the fraction will be less than 1, while when $[ion]_{high}$ is in the numerator, the fraction will be greater than 1. We want the configuration in which the fraction is less than 1 since $Q < K_{eq}$ the reaction will move towards equilibrium, and here, K_{eq} would be 1.

Suppose we have a concentration cell with 0.005 M Cu^{2+} in one cell and 0.10 M Cu^{2+} in another cell, we are asked to calculate the EMF of the concentration cell:

$$E = 0 - \frac{0.06V}{2} * \log\frac{0.005M}{0.1M} = -0.03 * \log\frac{5x10^{-3}}{1x10^{-1}} = -0.03 * \log(5*10^{-2})$$

$$E = -0.03(\log 5 + \log 10^{-2}) = -(0.03)(0.7-2) = -(0.03)(-1.3)$$

$$E = (1.3)(3*10^{-2}) = 3.9*10^{-2} \text{ or } 0.0392 \text{ V}$$

For any concentration cell, the standard state EMF $\left(E^\circ\right)$ is 0 since the two half-cells have the same half-reactions. Next, we substitute the concentration value of 0.005 M into the numerator and 0.1 M into the denominator since the smaller value goes in the numerator. We should also keep in mind that Cu is a 2+ ion, so the value of n, which is the moles of electrons, is 2. During our calculation, we can also use the following logarithmic rule: log(ab) = log a + log b. Ultimately, we get the value of 0.0392 V.

Logarithmic Approximations

One way of approaching logarithmic values without calculators is to use the below approximation. We can use the sequential odd numbers 3, 5, 7, and 9 to help us approximate the logarithmic values between 1 and 10, as in 3-5, 5-7, and 7-9 or log(3) - 0.5, log(5) - 0.7, and log(7) - 0.9.

Logarithm	Approximate Value	Actual Value
log 1	0	0
log 3	0.5	0.48
log 5	0.7	0.70
log 7	0.9	0.85
log 10	1.0	1.0

If we are given *log(0.073)* or *log(55)*, we should approach the calculation as follows:

$$log(0.0073) = log(7.3 x 10 \quad = 0.9 - 3.0 = -2.1$$

$$log(55) = log(5.5 x 10^1) = 0.7 + 1 = 1.7$$

Electrolytic Cells

Let's take a look at the final type of cell, the electrolytic cell. An electrolytic cell is an electrochemical cell in which the energy from an external power source is used to drive a normally non-spontaneous reaction, i.e. apply a reverse voltage to a voltaic cell. We encounter electrolytic cells during the charging phase of any type of rechargeable battery from the lead-acid battery in automobiles to the lithium-ion battery in smartphones.

In comparison to the galvanic cell, the electrodes of an electrolytic cell can be placed in a single compartment containing the molten or aqueous electrolyte. In addition, since the external battery source is what drives the electrons through the circuit, the electrodes will match the positive and negative terminal of the battery. While the anode remains the site of oxidation, it becomes the positive terminal, and the cathode becomes the negative terminal.

Another use for an electrolytic cell is for the decomposition of compounds, i.e. water, sodium chloride, into simpler compounds. Industrial processes take advantage of this in the production of chlorine or sodium hydroxide. Since electrolytic cells can be conducted in molten or aqueous electrolyte, depending on the cation and anion, the products from using a molten electrolyte may be different from the products from using an aqueous electrolyte.

In aqueous solution, at the cathode or negative electrode, if a metal is more reactive than hydrogen, then the reduced metal reacts with water to produce hydrogen gas. We can look to the activity series or a table of reduction potential to determine that the following cations are harder to reduce than water: Group 1A $\left(Li^+, Rb^+, k^+, Cs^+, Na^+\right)$ and Group llA $\left(Ba^{2+}, Sr^{2+}, Ca^{2+}, Mg^{2+}\right)$ cations. However, if a metal is less reactive than hydrogen, then the reduced metal does not react with water, i.e. copper, platinum, etc.

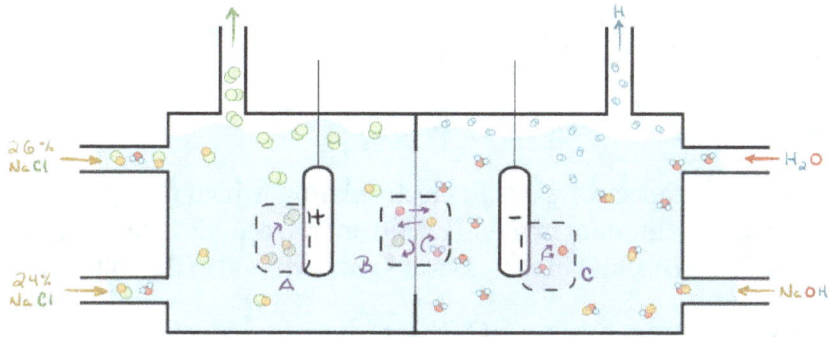

When looking at the electrolysis of sodium chloride in aqueous solution, there are additional reactions we must consider in calculating EMF. At the anode, the following reactions could occur:

$$2Cl^- \rightarrow Cl_2(g) + 2e^-; E^\circ = -1.36 \text{ V}$$
$$2H_2O \rightarrow O_2(g) + 4H^+ + 4e^-; E^\circ = -1.23 \text{ V}$$

While at the cathode, the following reactions could occur:

$$Na^+ + e^- \rightarrow Na(s); E^\circ = -2.7 \text{ V}$$
$$2H_2O + 2e^- \rightarrow H_2(g) + 2OH^-; Eo = -0.83 \text{ V}$$

At the anode, the oxidation of hydroxide is kinetically hindered because reactions involving oxygen gas are very slow. At the cathode, the reduction of water is more energetically favorable since it has a more positive reduction potential. NaOH is produced instead of sodium metal.

Suppose we are asked to calculate the EMF of the above electrolytic cell. We were given that the cell creates chlorine and sodium hydroxide, so we must pick these two reactions:

$$2Cl^- \rightarrow Cl_2(g) + 2e^-; E^\circ = -1.36 \text{ V}$$
$$2H_2O + 2e^- \rightarrow H_2(g) + 2OH^-; E^\circ = -0.83 \text{ V}$$

Normally we have to flip one of the equations into an oxidation potential, but the reaction is already written as an oxidation potential. We will most likely be given reduction potentials in a chart or table, but sometimes oxidation potentials are given. Adding the two EMF values together, the total E° for this electrolytic cell is -2.19 V, so it will require 2.19 V to drive this reaction.

In order to produce sodium or aluminum, we must use the electrolyte in molten state. For an electrolyte as molten sodium chloride, the following half-reactions take place:

$$Na + e^- \rightarrow Na(s); E = -2.7 \text{ V}$$
$$2Cl^- \rightarrow Cl(g) + 2e^-; E = -1.36 \text{ V}$$

In comparison to -2.19 V, the EMF for this cell is -4.0 V.

Electroplating and Faraday's Law

Another common use of electrolysis is in electroplating, which always occurs at the cathode. We can calculate the amount of metal plated on the cathode from the charge or current that passes through the cell, and in fact they are proportional in the following relationship:

$$Q \text{ (charge of } e^- \text{ from battery)} = Q \text{ (charge of } e^- \text{ for reduction)}$$

$$It = nF$$

where I is the current expressed in amperes or coulombs/second, n is the moles of electrons and F is Faraday's constant, which has the value of $96,485$ $C/mole^-$. We should always approximate Faraday's constant as 10^5 for our purposes in calculations.

Suppose we are asked to figure out the amount of platinum metal in grams that is deposited when 5.0 amps of current is passed through the electrolyte for 1 hour. Applying the equation, we plug in the following values:

$$(5.0\,A)(60\,min)\left(\frac{60\,sec}{1\,min}\right) = n\left(96,485\frac{C}{mol\,e^-}\right)$$

$$(5)(3600) = n(100000)$$

$$(18000) = n(100000)$$

$$n = 0.18 \text{ mol e}$$

$$(0.18 \text{ mol e})\left(\frac{1 \text{ mol}\,Pt}{2 \text{ mol e}}\right) = 0.09 \text{ mol}\,Pt$$

$$(0.09\,mol\,Pt)\left(195\,\frac{gPt}{mol\,Pt}\right) \approx (0.09\,mol\,Pt)\left(200\,\frac{gPt}{mol\,Pt}\right) = 18gPt$$

As expected, we approximated Faraday's constant as 100,000. We also have to convert 1 hour into seconds, which in total is 3600 seconds. The following math is straightforward and does not require much rounding. The only tricky part is remembering that n represents the mole of electrons. Since platinum forms a 2+ ion, then there are two moles of electrons consumed. To obtain the moles of platinum produced, we need to divide by 2. At the end, we can approximate 195 as 200 to obtain our final value of 18 grams.

We may expect to see 18.2 g, 36.4 g, 1.82 g and 3.64 g as possible answer choices to account for losing a power of 10 in our calculations or forgetting to divide by 2 to factor in that n represents the moles of electrons not the moles of the metal. Since there was little rounding, our calculated answer should pretty closely match one of the answer choices.

Coordination Chemistry

7

A coordination compound consists of a central metallic atom or ion surrounded by an array of ions or bound molecules. The study of coordination compounds is under the science of coordination chemistry. Some of the diverse topics covered in this chapter include coordination compounds, coordination complex, coordination number, trans and cis effect, etc.

Coordination chemistry emerged from the work of Alfred Werner, a Swiss chemist who examined different compounds composed of cobalt(III) chloride and ammonia. Upon the addition of hydrochloric acid, Werner observed that ammonia could not be completely removed. He then proposed that the ammonia must be bound more tightly to the central cobalt ion. However, when aqueous silver nitrate was added, one of the products formed was solid silver chloride. The amount of silver chloride formed was related to the number of ammonia molecules bound to the cobalt(III) chloride. For example, when silver nitrate was added to $CoCl_3 \cdot 6NH_3$ all three chlorides were converted to silver chloride. However, when silver nitrate was added to $CoCl_3 \cdot 5NH_3$, only 2 of the 3 chlorides formed silver chloride. When $CoCl_3 \cdot 4NH_3$ was treated with silver nitrate, one of the three chlorides precipitated as silver chloride.

The resulting observations suggested the formation of complex or coordination compounds. In the inner coordination sphere, which is also referred to in some texts as the first sphere, ligands are directly bound to the central metal. In the outer coordination sphere, sometimes referred to as the second sphere, other ions are attached to the complex ion. Werner was awarded the Nobel Prize in 1913 for his coordination theory. The following table is a summary of Werner's observations:

Initial compound	Resulting compounds upon adding $AgNO_3$
$CoCl_3 \cdot 6NH_3$	$\left[Co(NH_3)_6\right]^{3+}(Cl^-)_3$
$CoCl_3 \cdot 5NH_3$	$\left[Co(NH_3)_5 Cl\right]^{2+}(Cl^-)_2$
$CoCl_3 \cdot 4NH_3$	$\left[Co(NH_3)_4 Cl_2\right]^{+}(Cl^-)$
$CoCl_3 \cdot 3NH_3$	$\left[Co(NH_3)_3 Cl_3\right]$

As the table above shows, the complex ion $\left[Co(NH_3)_6\right]^{3+}$ is countered by the three chloride ions. The multi-level binding of coordination complexes play an important role in determining the dissociation of these complexes in aqueous solution. For example,

$[Co(NH_3)_5 Cl]^{2+}(Cl^-)_2$ dissociates into 3 ions while $[Co(NH_3)_4 Cl_2]^+(Cl^-)$ dissociates into 2 ions.ot further dissociate. By applying a current through the aqueous solutions of the resulting complex compounds, Werner measured the electrical conductivity and thus the dissociation properties of the complex compounds. The results confirmed his hypothesis of the formation of complex compounds. It is important to note that the above compounds have a coordination number of 6, which is a common coordination number for many inorganic complexes. Coordination numbers for complex compounds typically range from 1 to 16.

Properties of Coordination Complexes

Some methods of verifying the presence of complex ions include studying its chemical behavior. This can be achieved by observing the compounds' color, solubility, absorption spectrum, magnetic properties, etc. The properties of complex compounds are separate from the properties of the individual atoms. By forming coordination compounds, the properties of both the metal and the ligand are altered.

Metal-ligand bonds are typically thought of Lewis acid-base interactions. The metal atom acts as an electron pair acceptor (Lewis acid), while the ligands act as electron pair donors (Lewis base). The nature of the bond between metal and ligand is stronger than intermolecular forces because they form directional bonds between the metal ion and the ligand, but are weaker than covalent bonds and ionic bonds.

Common Ligands

Monodentate ligands donate one pair of electrons to the central metal atoms. An example of these ligands are the haldide ions (F^-, Cl^-, Br^-, I^-). Polydentate ligands, also called chelates or chelating agents, donate more than one pair of electrons to the metal atom forming a stronger bond and a more stable complex. A common chelating agent is ethylenediamine (en), which, as the name suggests, contains two ammines or: NH_2 sites which can bind to two sites on the central metal. An example of a tridentate ligand is bis-diethylenetriammine. An example of such a coordination complex is bis-diethylenetriamine cobalt[III].

Complex compound/ion	Coordination number	Oxidation State of Metal Atom
$[Fe(CN)_6]^{4-}$	6	2+
$[Co(NH_3)_4 SO_4]^-$	5	1+
$[Pt(NH_3)_4]^{2+}$	4	2+
$[Ni(NH_2CH_2CH_2NH_2)_3]^{2+}$	6	2+

Complex ions can form many compounds by binding with other complex ions in multiple ratios. This leads to many combinations of coordination compounds. The structures of certain coordination compounds can also have isomers, which can change their interactions with other chemical agents. The binding between metal and ligands is studied in metals, tetrahedral, and octahedral structures. There are many pharmaceutical and biological applications of coordination complexes and their isomers.

Coordination Compounds

Coordination compound is any of a class of substances with chemical structures in which a central metal atom is surrounded by nonmetal atoms or groups of atoms, called ligands, joined to it by chemical bonds.

Coordination compounds include such substances as vitamin B_{12}, hemoglobin, and chlorophyll, dyes and pigments, and catalysts used in preparing organic substances.

A major application of coordination compounds is their use as catalysts, which serve to alter the rate of chemical reactions. Certain complex metal catalysts, for example, play a key role in the production of polyethylene and polypropylene. In addition, a very stable class of organometallic coordination compounds has provided impetus to the development of organometallic chemistry. Organometallic coordination compounds are sometimes characterized by "sandwich" structures, in which two molecules of an unsaturated cyclic hydrocarbon, which lacks one or more hydrogen atoms, bond on either side of a metal atom. This results in a highly stable aromatic system.

Organometallic coordination compounds, which include transition metal compounds, may be characterized by "sandwich" structures that contain two unsaturated cyclic hydrocarbons on either side of a metal atom. Organometallic compounds are found in the

p-, d-, s-, and f- blocks of the periodic table (the purple-shaded blocks; the transition metals include those elements in the d- and f-blocks).

Coordination Compounds In Nature

Naturally occurring coordination compounds are vital to living organisms. Metal complexes play a variety of important roles in biological systems. Many enzymes, the naturally occurring catalysts that regulate biological processes, are metal complexes (metalloenzymes); for example, carboxypeptidase, a hydrolytic enzyme important in digestion, contains a zinc ion coordinated to several amino acid residues of the protein. Another enzyme, catalase, which is an efficient catalyst for the decomposition of hydrogen peroxide, contains iron-porphyrin complexes. In both cases, the coordinated metal ions are probably the sites of catalytic activity. Hemoglobin also contains iron-porphyrin complexes, its role as an oxygen carrier being related to the ability of the iron atoms to coordinate oxygen molecules reversibly. Other biologically important coordination compounds include chlorophyll (a magnesium-porphyrin complex) and vitamin B_{12}, a complex of cobalt with a macrocyclic ligand known as corrin.

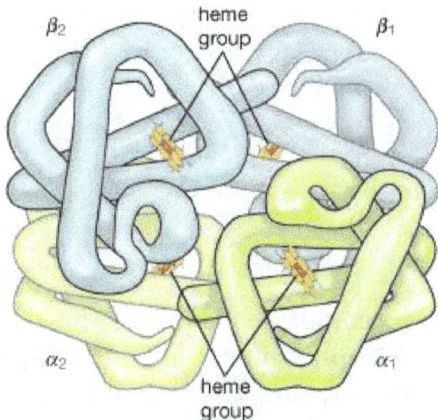

Hemoglobin is a protein made up of four polypeptide chains (α_1, α_2, β_1, and β_2). Each chain is attached to a heme group composed of porphyrin (an organic ringlike compound) attached to an iron atom. These iron-porphyrin complexes coordinate oxygen molecules reversibly, an ability directly related to the role of hemoglobin in oxygen transport in the blood.

Coordination Compounds In Industry

The applications of coordination compounds in chemistry and technology are many and varied. The brilliant and intense colours of many coordination compounds, such as Prussian blue, render them of great value as dyes and pigments. Phthalocyaninecomplexes (e.g., copper phthalocyanine), containing large-ring ligands closely related to the porphyrins, constitute an important class of dyes for fabrics.

Several important hydrometallurgical processes utilize metal complexes. Nickel, cobalt, and copper can be extracted from their ores as ammine complexes using aqueous ammonia. Differences in the stabilities and solubilities of the ammine complexes can be utilized in selective precipitation procedures that bring about separation of the metals. The purification of nickel can be effected by reaction with carbon monoxide to form the volatile tetracarbonylnickel complex, which can be distilled and thermally decomposed to deposit the pure metal. Aqueous cyanide solutions usually are employed to separate goldfrom its ores in the form of the extremely stable dicyanoaurate(−1) complex. Cyanide complexes also find application in electroplating.

There are a number of ways in which coordination compounds are used in the analysis of various substances. These include (1) the selective precipitation of metal ions as complexes—for example, nickel(2+) ion as the dimethylglyoxime complex (shown below):

(2) the formation of coloured complexes, such as the tetrachlorocobaltate(2−) ion, which can be determined spectrophotometrically—that is, by means of their light absorption properties, and (3) the preparation of complexes, such as metal acetylacetonates, which can be separated from aqueous solution by extraction with organic solvents.

In certain circumstances, the presence of metal ions is undesirable, as, for example, in water, in which calcium (Ca^{2+}) and magnesium (Mg^{2+}) ions cause hardness. In such cases the undesirable effects of the metal ions frequently can be eliminated by "sequestering" the ions as harmless complexes through the addition of an appropriate complexing reagent. Ethylenediaminetetraacetic acid (EDTA) forms very stable complexes, and it is widely used for this purpose. Its applications include water softening (by tying up Ca^{2+} and Mg^{2+}) and the preservation of organic substances, such as vegetable oils and rubber, in which case it combines with traces of transition metal ions that would catalyze oxidation of the organic substances.

A technological and scientific development of major significance was the discovery in 1954 that certain complex metal catalysts—namely, a combination of titanium trichloride, or TiCl$_3$, and triethylaluminum, or Al(C$_2$H$_5$)$_3$—bring about the polymerizations of organic compounds with carbon-carbon double bonds under mild conditions to form polymersof high molecular weight and highly ordered (stereoregular) structures. Certain of these polymers are of great commercial importance because they are used to make many kinds of fibres, films, and plastics. Other technologically important processes based on metal complex catalysts include the catalysis by metal carbonyls, such as hydridotetracarbonylcobalt, of the so-called hydroformylation of olefins—i.e., of their reactions with hydrogen and carbon monoxide to form aldehydes—and the catalysis by tetrachloropalladate(2–) ions of the oxidation of ethylene in aqueous solution to acetaldehyde.

Characteristics Of Coordination Compounds

Coordination compounds have been studied extensively because of what they reveal about molecular structure and chemical bonding, as well as because of the unusual chemical nature and useful properties of certain coordination compounds. The general class of coordination compounds or complexes, as they are sometimes called—is extensive and diverse. The substances in the class may be composed of electrically neutral molecules or of positively or negatively charged species (ions).

Among the many coordination compounds having neutral molecules is uranium (+6) fluoride, or uranium hexafluoride (UF$_6$). The structural formula of the compound represents the actual arrangement of atoms in the molecules:

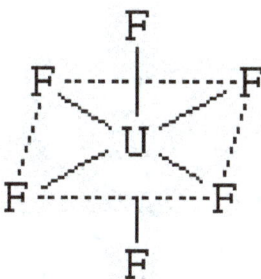

In this formula the solid lines, which represent bonds between atoms, show that four of the fluorine (F) atoms are bonded to the single atom of uranium (U) and lie in a plane with it, the plane being indicated by dotted lines (which do not represent bonds), whereas the remaining two fluorine atoms (also bonded to the uranium atom) lie above and below the plane, respectively.

An example of an ionic coordination complex is the hydrated ion of nickel, (Ni), hexaaquanickel (2+) ion, $\left[Ni(H_2O)_6\right]^{2+}$, the structure of which is shown below. In this structure, the symbols and lines are used as above, and the brackets and the "two plus" (2+) sign show that the double positive charge is assigned to the unit as a whole.

The central metal atom in a coordination compound itself may be neutral or charged (ionic). The coordinated groups—or ligands—may be neutral molecules such as water (in the above example), ammonia (NH_3), or carbon monoxide (CO); negatively charged ions (anions) such as the fluoride (in the first example above) or cyanide ion (CN^-); or, occasionally, positively charged ions (cations) such as the hydrazinium ($N_2H_5^+$) or nitrosonium (NO^+) ion.

Complex ions—that is, the ionic members of the family of coordination substances—may exist as free ions in solution, or they may be incorporated into crystalline materials (salts) with other ions of opposite charge. In such salts, the complex ion may be either the cationic (positively charged) or the anionic (negatively charged) component (or, on occasion, both). The hydrated nickel ion (above) is an example of a cationic complex. An anionic complex is the hexacyanide of the ferric iron (Fe^{+3}) ion, the hexacyanoferrate(3−) ion, $[Fe(CN)_6]^{3-}$, or

Crystalline salts containing complex ions include potassium hexacyanoferrate(3−) (potassium ferricyanide), $K_3[Fe(CN)_6]$, and the hexahydrate of nickel chloride, hexaaquanickel(2+) chloride, $[Ni(H_2O)_6]Cl_2$. In each case the charge on the complex ion is balanced by ions of opposite charge. In the case of potassium ferricyanide, three positively charged potassium ions, K^+, balance the negative charge on the complex, and in the nickel complex the positive charges are balanced by two negative chloride ions, Cl^-. The oxidation state of the central metal is determined from the charges on the ligands and the overall charge on the complex. For example, in hexaaquanickel(2+), water is electrically neutral and the charge on the complex ion is +2; thus, the oxidation state of Ni is +2. In hexacyanoferrate(3−), all six cyano ligands have a charge of −1; thus, the overall charge of −3 dictates that the oxidation state of Fe is +3.

The distinction between coordination compounds and other substances is, in fact, somewhat arbitrary. The designation coordination compound, however, is generally restricted to substances whose molecules or ions are discrete entities and in which the central atom is metal. Accordingly, molecules such as sulfur(+6) fluoride (sulfur hexa-fluoride; SF_6) and carbon(+4) fluoride (carbon tetrafluoride; CF_4) are not normally con-sidered coordination compounds, because sulfur (S) and carbon (C) are nonmetallic elements. Yet there is no great difference between these compounds and, say, uranium hexafluoride. Furthermore, such simple ionic salts as sodium chloride (NaCl) or nick-el(+2) fluoride (nickel difluoride; NiF_2) are not considered coordination compounds, because they consist of continuous ionic lattices rather than discrete molecules. Nev-ertheless, the arrangement (and bonding) of the anions surrounding the metal ions in these salts is similar to that in coordination compounds. Coordination compounds gen-erally display a variety of distinctive physical and chemical properties, such as colour, magnetic susceptibility, solubility and volatility, an ability to undergo oxidation-reduc-tion reactions, and catalytic activity.

A coordination compound is characterized by the nature of the central metal atom or ion, the oxidation state of the latter (that is, the gain or loss of electrons in passing from the neutral atom to the charged ion, sometimes referred to as the oxidation number), and the number, kind, and arrangement of the ligands. Because virtually all metallic elements form coordination compounds—sometimes in several oxidation states and usually with many different kinds of ligands—a large number of coordination compounds are known.

Coordination Number

Coordination number is the term proposed by Werner to denote the total number of bonds from the ligands to the metal atom. Coordination numbers generally range between 2 and 12, with 4 (tetracoordinate) and 6 (hexacoordinate) being the most com-mon. Werner referred to the central atom and the ligands surrounding it as the coordi-nation sphere. Coordination number should be distinguished from oxidation number (defined in the previous paragraph). The oxidation number, designated by an Arabic number with an appropriate sign (or, sometimes, by a Roman numeral in parentheses), is an index derived from a simple and formal set of rules and is not a direct indicator of electron distribution or of the charge on the central metal ion or compound as a whole. For the hexaamminecobalt(3+) ion, $[Co(NH_3)_6]^{3+}$, and the neutral moleculetriammine-trinitrocobalt(3+), $\left[Co(NO_2)_3(NH_3)_3\right]$, the coordination number of cobalt is 6 while its oxidation number is +3.

Ligands and Chelates

Each molecule or ion of a coordination compound includes a number of ligands, and, in any given substance, the ligands may be all alike, or they may be different. The term *ligand* was proposed by the German chemist Alfred Stock in 1916. Attachment of the ligands to the metal atom may be through only one atom, or it may be through several atoms. When only one atom is involved, the ligand is said to be monodentate; when two are involved, it is didentate, and so on. In general, ligands utilizing more than one bond are said to be polydentate. Because a polydentate ligand is joined to the metal atom in more than one place, the resulting complex is said to be cyclic—i.e., to contain a ring of atoms. Coordination compounds containing polydentate ligands are called chelates and their formation is termed *chelation*. Chelates are particularly stable and useful. An example of a typical chelate is bis(1,2-ethanediamine)copper(2+), the complex formed between the cupric ion (Cu^{2+}) and the organic compound ethylenediamine ($NH_2CH_2CH_2NH_2$, often abbreviated as en in formulas). The formula of the complex is

$$\left[Cu\left(NH_2CH_2CH_2NH_2\right)_2\right]^{2+}$$

and the structural formula is

$$\left[\begin{array}{cc} CH_2-H_2N & NH_2-CH_2 \\ | & | \\ & Cu \\ | & | \\ CH_2-H_2N & NH_2-CH_2 \end{array}\right]^{2+}$$

Mononuclear, Monodentate

The simplest types of coordination compounds are those containing a single metal atom or ion (mononuclear compounds) surrounded by monodentate ligands. Most of the coordination compounds already cited belong to this class. Among the ligands forming such complexes are a wide variety of neutral molecules (such as ammonia, water, carbon monoxide, and nitrogen), as well as monoatomic and polyatomic anions (such as the hydride, fluoride, chloride, oxide, hydroxide, nitrite, thiocyanate, carbonate, sulfate, and phosphate ions). Coordination of such ligands to the metal virtually always occurs through an atom possessing an unshared pair of electrons, which it donates to the metal to form a coordinate bond with the latter. Among the atoms that are known to coordinate to metals are those of virtually all the nonmetallic elements (such as hydrogen, carbon, oxygen, nitrogen, and sulfur), with the exception of the noble gases(helium [He], neon [Ne], argon [Ar], krypton [Kr], and xenon [Xe]).

Polydentate

The chelate complex of a copper ion and ethylenediamine mentioned above is an example of a compound formed between a metal ion and a didentate ligand. Two further examples of chelate complexes are shown below.

nickel-porphyrin complex

calcium-ethylenediamine-
tetraacetate complex

These are a nickel complex with a tetradentate large-ring ligand, known as a porphyrin, and a calcium complex with a hexadentate ligand, ethylenediaminetetraacetate (EDTA). Because metal-ligand attachment in such chelate complexes is through several bonds, such complexes tend to be very stable.

The commonest and most stable complexes of the lanthanoid metals (the series of 14 f-block elements following lanthanum [atomic number 57]) are those with chelating oxygen ligands, such as EDTA-type anions or hydroxo acids (e.g., tartaric or citric acids). The formation of such water-soluble complexes is employed in the separation of lanthanoids by ion-exchange chromatography. Lanthanoid β-diketonates are well known because some fluorinated β-diketonates yield volatile complexes amenable to gas-chromatographic separations. Neutral complexes can complex further to yield anionic species such as octacoordinated tetrakis(thenoyltrifluoroacetato)neodymate(1−), $[Nd(CF_3COCHCOCF_3)_4]^-$.

Certain ligands may be either monodentate or polydentate, depending on the particular compound in which they occur. The carbonate ion, $(CO_3)^{2-}$, for example, is coordinated to the cobalt (Co^{3+}) ions in two cobalt complexes, pentaamminecarbonatocobalt(+), $[Co(CO_3)(NH_3)_5]^+$, and tetraamminecarbonatocobalt(+), $[Co(CO_3)(NH_3)_4]^+$, through one and two oxygen atoms, respectively.

complex with metal-metal
bond (Re is rhenium)

complex with bridging ligands
(C_2H_5 is ethyl radical, Pt is platinum,
P is phosphorus, and Cl is chlorine)

complex with metal-metal
bonds and bridging ligands

complex with metal-cluster nucleus
○ = molybdenum atoms

◯ = chlorine atoms

Polynuclear

Polynuclear complexes are coordination compounds containing two or more metal atoms, or ions, in a single coordination sphere. The two atoms may be held together through direct metal-metal bonds, through bridging ligands, or both. Examples of each are shown above, along with a unique metal-cluster complex having six metal atoms in its nucleus.

Nomenclature

Generally, the systematic naming of coordination compounds is carried out by rules recommended by the International Union of Pure and Applied Chemistry (IUPAC). Among the more important of these are the following:

1. Neutral and cationic complexes are named by first identifying the ligands, followed by the metal; its oxidation number may be given in Roman numerals enclosed within parentheses. Alternatively, the overall charge on the complex may be given in Arabic numbers in parentheses. This convention is generally followed here. In formulas, anionic ligands (ending in -*o*; in general, if the anion name ends in -*ide*, -*ite*, or -*ate*, the final *e* is replaced by -*o*, giving -*ido*, -*ito*, and -*ato*) are cited in alphabetical order ahead of neutral ones also in alphabetical order (multiplicative prefixes are ignored). When the complex contains more than one ligand of a given kind, the number of such ligands is designated by one of the prefixes *di-*, *tri-*, *tetra-*, *penta-*, and so on or, in the case of complex ligands, by *bis-*, *tris-*, *tetrakis-*, *pentakis-*, and so on. In names (as opposed to formulas) the ligands are given in alphabetical order without regard to charge. The oxidation number of the metal is defined in the customary way as the residual charge on the metal if all the ligands were removed together with the electron pairs involved in coordination to the metal. The following examples are illustrative (aqua is the name of the water ligand):

    $\left[Ce\left(CH_3CCHO_2CCH_3\right)_4\right]$ tetrakis(2,4-pentanedionato) cerium(IV)

    $\left[Nd\left(H_2O\right)^9\right]^{3+}$ nonaaquaneodymium (III)

2. Anionic complexes are similarly named, except that the name is terminated by the suffix -*ate*; for example:

    $\left[Er(NCS)_6\right]^{3-}$ hexakis(thiocyanato) erbate(III)

3. In the case of salts, the cation is named first and then the anion; for example:

    $(NH_4)_2\left[Ce(NO_3)_6\right]$ ammonium hexanitatocerate(IV)

4. Polynuclear complexes are named as follows, bridging ligands being identified by a prefix consisting of the Greek letter *mu* (μ-):

$[(CO)_6 Mn - (CO)_5]$ decacarbonyldi-

                                          manganese (0) or

                                            bis (pentacarbonyl-

                                            manganese)

$[(NH_3)_5 Cr - OH - Cr(NH_3)_5]Cl_5$ m- hydroxo - bis[penta- ammine- chormium (III)]

                                            chloride

In addition to their systematic designations, many coordination compounds are also known by names reflecting their discoverers or colours. Examples are

$K[PtCl_3(C_2H_4)] \cdot H_2O$ Zeise's salt

$NH_4[Cr(SCN)_4(NH_3)_2] \cdot H_2O$ Reineoke's salt

$[Co(NH_3)_5(H_2O]Cl_3$ roseocobaltic chloride (red)

$KFe[Fe(CN)_6]$ Prussian blue

Structure And Bonding Of Coordination Compounds

Werner originally postulated that coordination compounds can be formed because the central atoms carry the capacity to form secondary, or coordinate, bonds, in addition to the normal, or valence, bonds. A more complete description of coordinate bonding, in terms of electron pairs, became possible in the 1920s, following the introduction of the concept that all covalent bonds consist of electron pairs shared between atoms, an idea advanced chiefly by the American physical chemist Gilbert N. Lewis. In Lewis's formulation, when both electrons are contributed by one of the atoms, as in the boron-nitrogen bond formed when the substance boron trifluoride (BF_3) combines with ammonia, the bond is called a coordinate bond:

boron   ammonia   compound with
trifluoride   coordinate bond

In Lewis's formulas, the valence (or bonding) electrons are indicated by dots, with each pair of dots between two atomic symbols representing a bond between the corresponding atoms.

Following Lewis's ideas, the suggestion was made that the bonds between metals and ligands were of this same type, with the ligands acting as electron donors and the metal ions as electron acceptors. This suggestion provided the first electronic interpretation of bonding in coordination compounds. The coordination reaction between silver ions and ammonia illustrates the resemblance of coordination compounds to the situation in the boron-nitrogen compound. According to this view, the metal ion can be regarded as a so-called Lewis acid and the ligands as Lewis bases:

$$
Ag^+ \;+\; 2 \;:\!\overset{\displaystyle H}{\underset{\displaystyle H}{N}}\!:H \;\longrightarrow\; \left[\; H\!:\!\overset{\displaystyle H}{\underset{\displaystyle H}{N}}\!:\!Ag\!:\!\overset{\displaystyle H}{\underset{\displaystyle H}{N}}\!:\!H \;\right]^{+}
$$

silver ion   ammonia   coordination complex

A coordinate bond may also be denoted by an arrow pointing from the donor to the acceptor.

Geometry

Many coordination compounds have distinct geometric structures. Two common forms are the square planar, in which four ligands are arranged at the corners of a hypothetical square around the central metal atom, and the octahedral, in which six ligands are arranged, four in a plane and one each above and below the plane. Altering the position of the ligands relative to one another can produce different compounds with the same chemical formula. Thus, a cobalt ion linked to two chloride ions and four molecules of ammonia can occur in both green and violet forms according to how the six ligands are placed. Replacing a ligand also can affect the colour. A cobalt ion linked to six ammonia molecules is yellow. Replacing one of the ammonia molecules with a water molecule turns it rose red. Replacing all six ammonia molecules with water molecules turns it purple.

Among the essential properties of coordination compounds are the number and arrangement of the ligands attached to the central metal atom or ion—that is, the coordination number and the coordination geometry, respectively. The coordination number of a particular complex is determined by the relative sizes of the metal atom and the ligands, by spatial (steric) constraints governing the shapes (conformations) of polydentate ligands, and by electronic factors, most notably the electronic configuration of the metal ion. Although coordination numbers from 1 to 16 are known, those below 3 and above 8 are rare. Possible structures and examples of species for the various coordination numbers are as follows: three, trigonal planar ($[Au\ \{P(C_6H_5)_3\}_3]^+$; four, tetrahedral ($[CoCl_4]^{2-}$) or square planar ($[PtCl_4]^{2-}$); five, trigonal bipyramid ($[CuCl_5]^{3-}$) or square pyramid ($VO(\text{acetylacetonate})_2$); six, octahedral ($[Co(NO_2)_6]^{3-}$) or trigonal prismatic ($[Re\ \{S_2C_2(C_6H_5)_2\}_3]$); seven, pentagonal bipyramid ($Na_5[Mo(CN)_7]10H_2O$), capped trigonal prism (cation in $[Ca(H_2O)_7]_2[Cd_6Cl_{16}(H_2O)_2]\ H_2O$), or capped octahedron (cation in $[Mo(CNC_6H_5)_7][PF_6]_2$); eight, square antiprism or dodecahedron ($[Zr(\text{acetylacetonate})_4]$; and nine, capped square antiprism ($La(NH_3)_9]^{3+}$) or tricapped trigonal prism ($[ReH_9]^{2-}$).

Coordination numbers are also affected by the 18-electron rule (sometimes called the noble gas rule), which states that coordination compounds in which the total number of valence electrons approaches but does not exceed 18 (the number of electrons in the valence shells of the noble gases) are most stable. The stabilities of 18-electron valence shells are also reflected in the coordination numbers of the stable mononuclear carbonyls of different metals that have oxidation number 0 e.g., tetracarbonylnickel, pentacarbonyliron, and hexacarbonylchromium (each of which has a valence shell of 18).

The 18-electron rule applies particularly to covalent complexes, such as the cyanides, carbonyls, and phosphines. For more ionic (also called outer-orbital) complexes, such as fluoro or aqua complexes, electronic factors are less important in determining coordination numbers, and configurations corresponding to more than 18 valence electrons are not uncommon. Several nickel (+2) complexes, for example including the hexafluoro, hexaaqua, and hexaammine complexes—each have 20 valence electrons.

Any one metal ion tends to have the same coordination number in different complexes e.g., generally six for chromium (+3) but this is not invariably so. Differences in coordination number may result from differences in the sizes of the ligands; for example, the iron (+3) ion is able to accommodate six fluoride ions in the hexafluoro complex $[FeF_6]^{3-}$ but only four of the larger chloride ions in the tetrachloro complex $[FeCl_4]^-$. In some cases, a metal ion and a ligand form two or more complexes with different coordination numbers e.g., tetracyanonickelate $[Ni(CN)_4]^{2-}$ and pentacyanonickelate $[Ni(CN)_5]^{3-}$, both of which contain Ni in the +2 oxidation state.

Isomerism

Coordination compounds often exist as isomers-i.e., as compounds with the same chemical composition but different structural formulas. Many different kinds of isomerism occur among coordination compounds. The following are some of the more common types.

Cis-trans Isomerism

Cis-trans (geometric) isomers of coordination compounds differ from one another only in the manner in which the ligands are distributed spatially; for example, in the isomeric pair of diamminedichloroplatinum compounds

the two ammonia molecules and the two chlorine atoms are situated next to one another in one isomer, called the *cis* (Latin for "on this side") isomer, and across from one another in the other, the *trans* (Latin for "on the other side") isomer. A similar relationship exists between the *cis* and *trans* forms of the tetraamminedichlorocobalt(1+) ion:

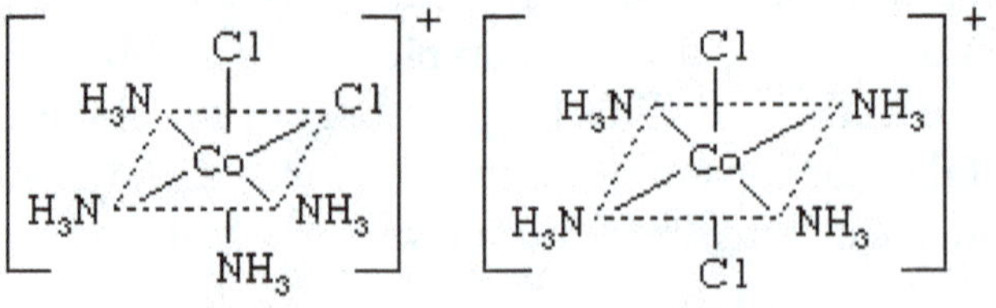

$$cis \qquad\qquad\qquad trans$$

Enantiomers and Diastereomers

So-called optical isomers (or enantiomers) have the ability to rotate plane-polarized light in opposite directions. Enantiomers exist when the molecules of the substances are mirror images but are not superimposable upon one another. In coordination compounds, enantiomers can arise either from the presence of an asymmetric ligand, such as one isomer of the amino acid, alanine (aminopropionic acid),

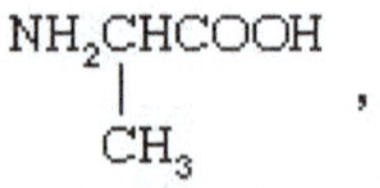

$$\begin{array}{c} NH_2CHCOOH \\ | \\ CH_3 \end{array} \quad ,$$

or from an asymmetric arrangement of the ligands. Familiar examples of the latter variety are octahedral complexes carrying three didentate ligands, such as ethylenediamine, $NH_2CH_2CH_2NH_2$. The two enantiomers corresponding to such a complex are depicted by the structures below.

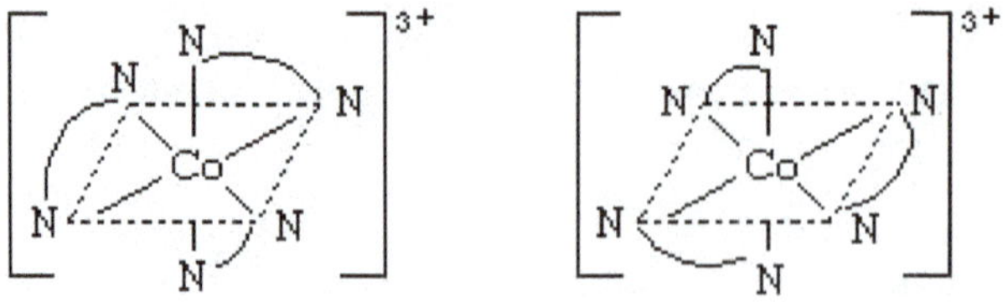

The ethylenediamine ligands above are indicated by a curved line between the symbols for the nitrogen atoms.

Diastereomers, on the other hand, are not superimposable and also are not mirror images. Using AB as an example of a chelating ligand, in which the symbol AB implies that the two ends of the chelate are different, there are six possible isomers of a complex $cis\text{-}[M(AB)_2 X_2]$. For example, AB might correspond to alanine $[CH_3CH(NH_2)C(O)O]^-$, where both N and O are attached to the metal. Alternatively, AB could represent a ligand such as propylenenediamine, $[NH_2CH_2C(CH_3)HNH_2]$, where the two ends of the molecule are distinguished by the fact that one of the Hs on a C is substituted with a methyl (CH_3) group.

Ionization Isomerism

Certain isomeric pairs occur that differ only in that two ionic groups exchange positions within (and without) the primary coordination sphere. These are called ionization isomers and are exemplified by the two compounds, pentaamminebromocobalt sulfate, $[CoBr(NH_3)_5]SO_4$, and pentaamminesulfatocobalt bromide, $[Co(SO_4)(NH_3)_5]$ Br. In the former the bromide ion is coordinated to the cobalt (3+) ion, and the sulfate ion is outside the coordination sphere; in the latter the sulfate ion occurs within the coordination sphere, and the bromide ion is outside it.

Linkage Isomerism

Isomerism also results when a given ligand is joined to the central atom through different atoms of the ligand. Such isomerism is called linkage isomerism. A pair of linkage isomers are the ions $[Co(NO_2)(NH_3)_5]^{2+}$ and $[Co(ONO)(NH_3)_5]^{2+}$, in which the anionic ligand is joined to the cobalt atom through nitrogen or oxygen, as shown by designating it with the formulas NO_2^-(nitro) and ONO^-(nitrito), respectively. Another example of this variety of isomerism is given by the pair of ions $[Co(CN)_5(NCS)]^{3-}$ and $[Co(CN)_5(SCN)]^{3-}$, in which an isothiocyanate $(NCS)^-$ and a thiocyanate group $(SCN)^-$ are bonded to the cobalt(3+) ion through a nitrogen or sulfur atom, respectively.

Coordination Isomerism

Ionic coordination compounds that contain complex cations and anionscan exist as isomers if the ligands associated with the two metal atoms are exchanged, as in the pair of compounds, hexaamminecobalt(3+) hexacyanochromate(3−), $[Co(NH_3)_6][Cr(CN)_6]$, and hexaamminechromium(3+) hexacyanocobaltate(3−), $[Cr(NH_3)_6][Co(CN)_6]$. Such compounds are called coordination isomers, as are the isomeric pairs obtained by redistributing the ligands between the two metal atoms, as in the doubly coordinated pair, tetraammineplatinum(2+) hexachloroplatinate(2−), $[Pt(NH_3)_4][PtCl_6]$, and tetraamminedichloroplatinum(2+) tetrachloroplatinate(2−), $[PtCl_2(NH_3)_4][PtCl_4]$.

Ligand Isomerism

Isomeric coordination compounds are known in which the overall isomerism results from isomerism solely within the ligand groups. An example of such isomerism is shown by the ions, bis(1,3-diaminopropane)platinum(2+) and bis(1,2-diaminopropane) platinum(2+),

$$\left[Pt\left(NH_2CH_2CH_2CH_2NH_2\right)_2\right]^{2+} \text{ and } \left[Pt\left(NH_2\underset{|}{C}HCH_2NH_2\right)_2\right]^{2+}.$$
$$CH$$

Important Types of Reactions Of Coordination Compounds

Acid-base

Coordination to a positive metal ion usually enhances the acidity (i.e., the tendency to lose protons) of hydrogen-containing ligands, such as water and ammonia. Thus, many metal ions in aqueous solution commonly exhibit acidic behaviour. Such behaviour is exemplified by hydrolysis reactions of the type shown in the following equilibrium:

$$[M(H_2O)_x]^{n+} \rightleftharpoons [M(OH)(H_2O)_{x-1}]^{(n-1)+} + H^+,$$

in which M represents the metal ion, n its charge, and x the number of coordinated water molecules.

The acidities of such aqua ions depend on the charge, size, and electronic configuration of the metal ion. This dependence is reflected in the values of acid dissociation constants, which range from about 10^{-14} (a value only slightly larger than for pure water, for which the dissociation constant = $10^{-15.7}$) for the hydrated lithium ion, to about 10^{-2} (a value equivalent to that of a fairly strong acid) for the hydrated uranium(4+) ion. Acid-base equilibria are rapidly established in solution, generally within a fraction of a second.

In some cases, hydrolysis of a metal ion may be accompanied by polymerization to form dinuclear or polynuclear hydroxo- or oxygen-bridged complexes.

$$2[Fe(H_2O)_6]^{3+} \rightleftharpoons \left[(H_2O)_4Fe \underset{O}{\overset{O}{\diamond}} Fe(H_2O)_4 \right]^{4+} + 2H^+ + 2H_2O.$$

Even very weakly acidic ligands, such as ammonia, can acquire appreciable acidity through coordination to a metal ion. Thus, the hexaammineplatinum(4+) ion dissociates according to the following equilibrium:

$$[Pt(NH_3)_6]^{4+} \rightleftharpoons [Pt(NH_2)(NH_3)_5]^{3+} + H^+$$

In addition to intrinsic strength, acids and bases have other properties that determine the extent of reactions. According to the hard and soft acids and bases (HSAB) theory, the metal cation and anion are considered to be acids and bases, respectively. Hard acids and bases are small and nonpolarizable, whereas soft acids and bases are larger and more polarizable. Interactions between two hard or soft acids or bases are stronger than ones between one hard and one soft acid or base. The theory can be used to explain solubilities, formation of metallic ores, and some reactions of metal cations with ligands.

Substitution

One of the most general reactions exhibited by coordination compounds is that of substitution, or replacement, of one ligand by another. This process is depicted in a generalized manner by the equation $ML_{x-1}Y + Z \rightarrow ML_{x-1}Z + Y$ for a metal complex of coordination number x. The ligands L, Y, and Z may be chemically similar or different. (Charges have been omitted here for simplicity.)

A class of substitution reactions that affords the widest possible comparison of different metal ions is the replacement of water in the coordination spheres of metal-aqua complexes in aqueous solution. The substitution may be by another water molecule (which can be labeled with the isotope oxygen-18 to permit the reaction to be followed) or by a different ligand, such as the chloride ion. Reactions of both types occur as shown below (oxygen-18 is indicated by the symbol $\overset{*}{O}$).

$$[M(H_2O_x)]^{n+} + H_2\overset{*}{O} \rightarrow [M(H_2O)_{x-1}(H_2\overset{*}{O})]^{n+} + H_2O$$

$$[M(H_2O_x)]^{n+} + Cl^- \rightarrow [M(H_2O)_{x-1}Cl]^{(n-1)+} + H_2O$$

Many such reactions are extremely fast, and it has been only since 1950, following the development of appropriate experimental methods (including stopped flow, nuclear magnetic resonance, and relaxation spectrometry), that the kinetics and mechanisms of this class of reactions have been extensively investigated. Rates of substitution of metal-aqua ions have been found to span a wide range, the characteristic times required for substitution ranging from less than 10^{-9} second for monopositive ions, such as hydrated potassium ions, to several days for certain more highly charged ions, such as hexaaquachromium(3+) and hexaaquarhodium(3+). The rate of substitution parallels the ease of loss of a water molecule from the coordination sphere of the aqua complex and thus increases with increasing size and with decreasing charge of the metal ion. For transition metal ions, electronic factors also have an important influence on rates of substitution.

There are two limiting mechanisms (or pathways) through which substitution may occur—namely, dissociative and associative mechanisms. In the dissociative mechanism, a ligand is lost from the complex to give an intermediate compound of lower coordination number. This type of reaction path is typical of octahedral complexes, many aqua complexes, and metal carbonyls such as tetracarbonylnickel. An example of a dissociative reaction pathway for an octahedral complex of cobalt is as follows:

$$[Co(CN)_5(H_2O)]^{2-} \xrightarrow{-H_2O} [Co(CH)_5]^2 \xrightarrow{+I^-} [Co(CN)_5 I]^{3-}.$$

The associative mechanism for substitution reactions, on the other hand, involves association of an extra ligand with the complex to give an intermediate of higher coordination number; one of the original ligands is then lost to restore the initial coordination number. Substitution reactions of square planar complexes, such as those of the nickel(2+), palladium(2+), and platinum(2+) ions, usually proceed through associative pathways involving intermediates with coordination number five. An example of a reaction following such a pathway is

$$[Pt(NH_3)_4]^{2+} \xrightarrow{\; Cl^- \;} \left[(NH_3)_3Pt \begin{matrix} NH_3 \\ Cl \end{matrix} \right]^+ \xrightarrow{\; -NH_3 \;} [Pt(NH_3)_3Cl]^+$$

A characteristic feature of this class of reactions is the sensitivity of the rate of substitution of a given ligand to the nature of the ligand in the *trans* position. The *trans* ligand activates a ligand for replacement as follows, in decreasing order:

$$CO, CN^-, C_2H_4 > PR_3, H^- > NO_2^-, I^-, SCN^- > Br^-, Cl^- > NH_3, H_2O.$$

The *trans* effect may be used for synthetic purposes; thus, the reaction of the tetrachloroplatinate(2−) ion with ammonia yields *cis*-diamminedichloroplatinum, whereas the reaction of the tetraammineplatinum(2+) ion with the chloride ion gives the *trans* isomer, *trans*-diamminedichloroplatinum. The reactions are shown below.

$$\left[\begin{matrix} & Cl & \\ Cl - & Pt & - Cl \\ & Cl & \end{matrix} \right]^{2-} \xrightarrow[-Cl^-]{+NH_3} \left[\begin{matrix} & Cl & \\ Cl - & Pt & - NH_3 \\ & Cl & \end{matrix} \right] \xrightarrow[-Cl^-]{+NH_3} \left[\begin{matrix} & Cl & \\ Cl - & Pt & - NH_3 \\ & NH_3 & \end{matrix} \right]$$

$$\left[\begin{matrix} & NH_3 & \\ H_3N - & Pt & - NH_3 \\ & NH_3 & \end{matrix} \right]^{2+} \xrightarrow[-NH_3]{+Cl^-} \left[\begin{matrix} & NH_3 & \\ H_3N - & Pt & - Cl \\ & NH_3 & \end{matrix} \right]^+ \xrightarrow[-NH_3]{+Cl^-} \left[\begin{matrix} & NH_3 & \\ Cl - & Pt & - Cl \\ & NH_3 & \end{matrix} \right]$$

In both reactions, the *trans* effect causes the introduction of the ligand *trans* to chloride rather than *trans* to ammonia.

Lability and Inertness

In considering the mechanisms of substitution (exchange) reactions, Canadian-born American chemist Henry Taube distinguished between complexes that are labile (reacting completely in about one minute in 0.1 M solution at room temperature [25 °C, or 77 °F]) and those that are inert (under the same conditions, reacting either too slowly to measure or slowly enough to be followed by conventional techniques). These terms refer to kinetics (reaction rates) and should not be confused with the thermodynamic terms unstable and stable, which refer to equilibrium. For example, as mentioned above, most cyanide complexes are extremely stable (they possess very small dissociation constants); yet, if their rate of exchange with carbon-14-labeled cyanide, as represented in the following equation,

$$[M(CN)_x]^{y-} + x^{14}CN^- \rightleftharpoons [M(^{14}CN)_x]^{y-} + xCN^-,$$

is measured, $[Ni(CN)_4]^{2-}$ and $[Hg(CN)_4]^{2-}$ are found to be labile, whereas $[Mn(CN)_6]^{3-}, [Fe(CN)_6]^{4-}, [Fe(CN)_6]^{3-}$, and $[Cr(CN)_6]^{3-}$ are inert. On the other hand, $[Co(NH_3)_6]^{3+}$, a kinetically inert complex, is thermodynamically stable in acidic solution. Inertness may result from the lack of a suitable low-energy pathway for the reaction. In short, stable complexes possess large positive free energies of reaction (ΔG), whereas inert complexes merely possess large positive free energies of activation (ΔG^*).

While the existence of geometric or optical isomers in the solid state or in solution at nonequilibrium concentrations is evidence supporting the inertness of the complex, this does not constitute absolute proof. Conversely, the possibility of intramolecular rearrangement means that failure to isolate geometric isomers or to resolve the racemic mixture into optical isomers is not absolute proof of lability.

Taube has interpreted lability of complexes according to their electronic configuration in terms of VB theory. Labile complexes are either of the outer orbital type (outer d orbitals involved in bonding—e.g., sp^3d^2 as opposed to d^2sp^3 [inner orbital] for octahedral complexes) or of the inner orbital type with at least one vacant d orbital (available for accommodation of a seventh group during the [associative] substitution reaction).

Isomerization

Coordination compounds that exist in two or more isomeric forms may undergo reactions that convert one isomer to another. Examples are the linkage isomerization and *cis-trans* isomerization reactions depicted below.

$$[Co(ONO)(NH_3)_5]^{2+} \rightarrow [Co(NO_2)(NH_3)_5]^{2+}$$
$$cis\text{-}[Co(H_2O)_2(NH_2CH_2CH_2NH_2)_2]^3 \rightarrow$$
$$trans\text{-}[Co(H_2O)_2(NH_2CH_2CH_2NH_2)_2]^{3+}$$

The first of these has been shown to proceed intramolecularly (i.e., without dissociation of the nitrite ligand), whereas the second probably occurs through dissociation of one of the water-molecule ligands.

Oxidation-reduction

Transition metals commonly exhibit two or more stable oxidation states, and their complexes accordingly are able to undergo oxidation-reduction reactions. The simplest such reactions involve electron transfer between two complexes, with little if any accompanying rearrangement or chemical change. An example is shown below:

$$[Fe(CN)_6]^{4-} + [IrCl_6]^{2-} \rightarrow [Fe(CN)_6]^{3-} + [IrCl_6]^{3-} \,.$$

In other cases, oxidation-reduction is accompanied by significant chemical rearrangement. An example is

$$[CoCl(NH_3)_5]^{2+} + [Cr(H_2O)_6]^{2+} + 5[H_3O]^+ \rightarrow$$
$$[Co(H_2O)_6]^{2+} + [CrCl(H_2O)_5]^{2+} + 5[NH_4]^+ \,.$$

Two limiting mechanisms of electron transfer, commonly designated outer-sphere and inner-sphere mechanisms, have been recognized. Outer-sphere electron transfer occurs without dissociation or disruption of the coordination sphere of either complex—i.e., through both intact coordination spheres. The first reaction above is of this type. On the other hand, inner-sphere electron transfer—e.g., the second reaction above— proceeds by formation of a dinuclear complex in which the two metal ions are joined by a common bridging ligand (in this case the chloride ion) through which the electron is transferred. Such electron transfer also may occur through polyatomic bridging ligands to which the two metal ions are attached at different sites separated by several atoms, as in the reduction of pentaammine(isonicotinamide)cobalt(3+) by hexaaquachromium(2+) ion through a bridged intermediate:

Strikingly large differences in rates of electron transfer are observed even between closely related reactions. Thus, the rate of reduction of the pentaamminebromocobalt(3+) ion by the hexaaquachromium(2+) ion is about 10^7 times higher than that of the acetatopentaamminecobalt(2+) ion by the same chromium ion.

Synthesis of Coordination Compounds

The great variety of coordination compounds is matched by the diversity of methods through which such compounds can be synthesized. Complex halides, for example, may be prepared by direct combination of two halide salts (either in the molten state or in a suitable solvent). Palladium chloride and potassium chloride, for example, react to give the complex potassium tetrachloropalladate(2−), as shown in the following equation:

$$PdCl_2 + 2KCl \rightarrow K_2[PdCl_4]$$

Another widely used route to coordination compounds is through the direct combination of a metal ion and appropriate ligands in solution. Thus, the addition of a sufficiently high concentration of ammonia to an aqueous solution of a nickel(2+) salt leads, through a series of reactions, to the formation of the hexaamminenickel(2+) ion, which can be precipitated, for example, as the sulfate salt, $[Ni(NH_3)_6]SO_4$.

Complexes of metal ions in high oxidation states are sometimes more readily formed by adding the ligands to a solution of the metal ion in a lower oxidation state in the presence of an oxidizing agent. Thus, addition of ammonia to an aqueous solution of a cobalt(2+) salt in the presence of air or oxygen leads to the formation of cobalt(3+)-ammine complexes such as hexaamminecobalt(3+), $[Co(NH_3)_6]^{3+}$, and pentaamminea-quacobalt(3+), $[Co(NH_3)_5(H_2O)]^{3+}$, ions.

Complexes of metals in low oxidation states, such as the carbonyls of metals in their zero oxidation states, can sometimes be prepared by direct combination of the metal with the ligand, as, for example, in the reaction of nickel metal with carbon monoxide.

$$Ni(solid) + 4CO \xrightarrow[\text{1 atm}]{25°} Ni(CO)_4$$

More commonly, a salt of the metal is reduced in the presence of the ligand. An example of this type of synthesis is the reduction of cobalt carbonate with hydrogen in the presence of carbon monoxide to give bis(tetracarbonylcobalt).

$$2CoCO_3 + 2H_2 + 8CO \xrightarrow[\text{250 – 300 atm}]{120 - 200^*} Co_2(CO)_8 + 2CO_2 + 2H_2O$$

Similar procedures are applicable to the synthesis of metal sandwich compounds containing cyclopentadienyl and benzene ligands. Dibenzenechromium, for example, can be prepared from chromic chloride, benzene, and aluminum, as shown in the following equation.

$$CrCl_3 + 2C_6H_6 + Al \rightarrow Cr(C_6H_6)_2 + AlCl_3$$

Hydrido complexes of transition metals can be prepared by reactions of suitable precursors either with molecular hydrogen or with suitable reducing agents such as hydrazine or sodium borohydride; for example,

$$2[Co(CN)_5]^{3-} + H_2 \rightarrow 2[CoH(CN)_5]^{3-}.$$

Transition metal complexes containing metal-carbon bonds can be prepared by a variety of routes, some of the more important of which are illustrated by the following examples.

$$Mn(CO)_5Br + LiCH_3 \rightarrow Mn(CO)_5(CH_3) + LiBr,$$
$$[Mn(CO)_5]^- + CH_3Br \rightarrow Mn(CO)_5(CH_3) + Br^-,$$
$$ptHCl[P(CH_3CH_2)_3]_2 + CH_2 = CH_2 \rightarrow$$
$$pt(CH_2CH_3)Cl[p(CH_3CH_2)_3]_2 , \text{ and}$$

$$2[Co(CN)_5]^{3-} + CH_3I \rightarrow$$

$$[Co(CN)_5(CH_3)]^{3-} + [Co(CN)_5 I]^{3-}$$

Coordination Complex

Coordination complexes are formed between a metal ion (Lewis acid) and ligands (Lewis base). The splitting of the d-orbitals (crystal field splitting) and the distribution of the d-electrons of the central transition metal ion is dependent upon the ligands that bind to the metal center. The separation of the d-orbitals is called the crystal field splitting energy and is within the visible region of the electromagnetic spectrum (400 nm – 750 nm).

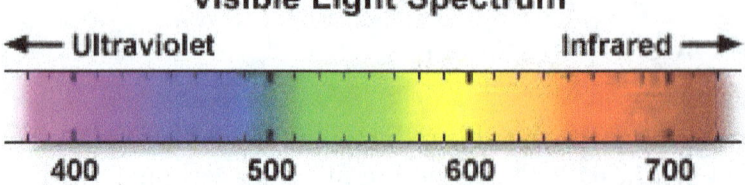

Transition metal complexes have a very wide range of colors within the visible spectrum thus they are commonly found as ingredients in colored paint. A transition metal complex will absorb a photon of light which will excite an electron into a higher energy orbit. The electron then returns to the ground-state emitting a photon within the visible region of the electromagnetic spectrum. The energy of the absorbed light may be calculated as follows:

$$\Delta = E = h\nu = \frac{hc}{\lambda}$$

Spectroscopy is a method of analysis based on measuring the energy of light absorbed by a substance and relating that energy to structural characteristics. When a complex absorbs white light, the light left over is the observed color or complementary color of the coordination complex. A color wheel may be used to determine complementary colors.

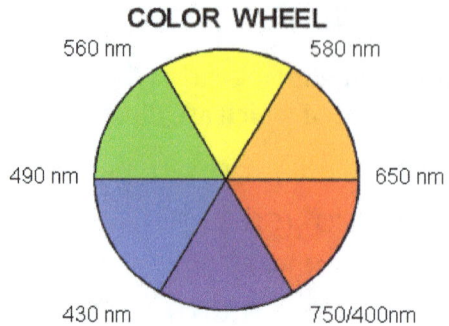

Principal Types Of Complexes

The tendency for complexes to form between a metal ion and a particular combination of ligands and the properties of the resulting complexes depend on a variety of properties of both the metal ion and the ligands. Among the pertinent properties of the metal ion are its size, charge, and electron configuration. Relevant properties of the ligand include its size and charge, the number and kinds of atoms available for coordination, the sizes of the resulting chelate rings formed (if any), and a variety of other geometric (steric) and electronic factors.

Many elements, notably certain metals, exhibit a range of oxidation states—that is, they are able to gain or lose varying numbers of electrons. The relative stabilities of these oxidation states are markedly affected by coordination of different ligands. The highest oxidation states correspond to empty or nearly empty d subshells (as the patterns of $d$orbitals are called). These states are generally stabilized most effectively by small negative ligands, such as fluorine and oxygen atoms, which possess unshared electron pairs. Such stabilization reflects, in part, the contribution of π bonding caused by electron donation from the ligands to empty d orbitals of the metal ions in the complexes. Conversely, neutral ligands, such as carbon monoxide and unsaturated hydrocarbons, which are relatively poor electron donors but which can accept π electrons from filled d orbitals of the metal, tend to stabilize the lowest oxidation states of metals. Intermediate oxidation states are most effectively stabilized by ligands such as water, ammonia, and cyanide ion, which are moderately good σ–electron donors but relatively poor π–electron donors or acceptors.

Chromium complexes of various oxidation states		
oxidation state	electron configuration*	coordination complex
+6	d^0	$[CrF_6], [CrO_4]^{2-}$
+5	d^1	$[CrO_4]^{3-}$
+4	d^2	$[CrO_4]^{4-}, [Cr(OR)_4]$**
+3	d^3	$[Cr(H_2O)_6]^{3+}, [Cr(NH_3)_6]^{3+}$
+2	d^4	$[Cr(H_2O)_6]^{2+}$
0	d^6	$[Cr(CO)_6], [Cr(C_6H_6)_2]$
**R symbolizes an organic alkyl radical. *Number of d electrons indicated by superscript.		

Aqua Complexes

Few ligands equal water with respect to the number and variety of metal ions with which they form complexes. Nearly all metallic elements form aqua complexes, frequently in more than one oxidation state. Such aqua complexes include hydrated ions in aqueous solution as well as hydrated salts such as hexaaquachromium(3+) chloride, $[Cr(H_2O)_6]$ Cl_3. For metal ions with partially filled d subshells (i.e., transition metals), the coordination numbers and geometries of the hydrated ions in solution can be inferred from their light-absorption spectra, which are generally consistent with octahedral coordination by six water molecules. Higher coordination numbers probably occur for the hydrated rare-earth ions such as lanthanum(3+).

When other ligands are added to an aqueous solution of a metal ion, replacement of water molecules in the coordination sphere may occur, with the resultant formation of other complexes. Such replacement is generally a stepwise process, as illustrated by the following series of reactions that results from the progressive addition of ammonia to an aqueous solution of a nickel(2+) salt:

$$[Ni(H_2O)_6]^{2+} + NH_3 \rightleftharpoons [Ni(NH_3)(H_2O)_5]^{2+} + H_2O$$

With increasing additions of ammonia, the equilibria are shifted toward the higher ammine complexes (those with more ammonia and less water) until ultimately the hexaamminenickel(2+) ion predominates:

$$[Ni(NH_3)_5(H_2O)]^{2+} + NH_3 \rightleftharpoons [Ni(NH_3)_6]^{2+} + H_2O$$

The tendency of metal ions in aqueous solution to form complexes with ammonia as well as with organic amines (derivatives of ammonia, with chains of carbon atoms attached to the nitrogen atom) is widespread. The stabilities of such complexes exhibit a considerable range of dependence on the nature of the metal ion as well as on that of the amine. The marked enhancement of stability that results from chelation is reflected in the equilibrium constants of the reactions values that indicate the relative proportions of the starting materials and the products at equilibrium. Complexes of hexaaquanickel(2+) ions can be formed with a series of polyamines i.e.,

$$[Ni(H_2O)_6]^{2+} + nL \rightleftharpoons [NiL_n(H_2O)_{6-n}]^{2+} + nH_2O,$$

in which L is the ligand and n the number of water molecules displaced from the complex. In this series the equilibrium constants, K_L, increase dramatically as the possibilities for chelation increase (that is, as the number of nitrogen atoms available for bonding to the metal atom increases).

It should be noted that, in the particular examples cited above, the coordination number of the metal ion is invariant throughout the substitution process, but this is not always the case. Thus, the ultimate products of the addition of the cyanide ion to an aqueous solution of hexaaquanickel(2+) ion are tetracyanonickelate(2−) and pentacyanonickelate(3−), both containing nickel in the +2 oxidation state. Similarly, addition of the chloride ion to a solution of hexaaquairon(3+) yields tetrachloroferrate(3−). Both complexes contain iron in the same oxidation state of +3.

Halo Complexes

Probably the most widespread class of complexes involving anionic ligands is that of the complexes of the halide ions—i.e., the fluoride, chloride, bromide, and iodide ions. In addition to forming simple halide salts, such as sodium chloride and nickel difluoride (in which the metal ions are surrounded by halide ions, these in a sense being regarded as coordinated to them), many metals form complex halide salts—such as potassium tetrachloroplatinate(2−), $K_2[PtCl_4]$—that contain discrete complex ions. Most metal ions also form halide complexes in aqueous solution. The stabilities of such complexes span an enormous range—from the alkali-metal ions (lithium, sodium, potassium, and so on), whose formation of halide complexes in aqueous solution can barely be detected, to extremely stable halide complexes, such as the tetraiodomercurate(2−), tetrachlorothallate(1−), and tetrachloropalladate(2−) ions, the extent of whose dissociation is extremely small.

The stabilities of halide complexes reflect a pattern by which metal ions can be divided into two general classes, designated as A and B or as hard and soft, respectively. (Generally, the electrons in the atoms of the hard elements are considered to form a compact and not easily deformable group, whereas those in the atoms of the soft elements form a looser group—that is, one more easily deformed.) For the former class which includes Be, Mg, Sc, Cr, Fe, Ni, Cu, In, and Sn the order of increasing stability of the halide complexes in aqueous solution is iodides < bromides < chlorides < fluorides. Conversely, for the class B (or soft) ions such as Pt, Ag, Cd, Hg, Tl, and Pb the order of increasing stability of the halide complexes is fluorides < chlorides < bromides < iodides. In contrast to class-A metals, those of class B also tend to form more stable complexes with sulfur-containing ligands than with oxygen-containing ligands and more stable complexes with phosphorus ligands than with nitrogen ligands.

Carbonyl Complexes

Following the discovery of the first metal carbonyl complex, tetracarbonylnickel, Ni(-CO)$_4$, in 1890, many compounds containing carbon monoxide coordinated to transition metals have been prepared and characterized. For reasons already discussed, such compounds generally contain metal atoms or ions in low oxidation states. The following are some of the more common types of metal carbonyl compounds: (1) simple mononuclear carbonyls of metals in the zero oxidation state, such as tetracarbonylnickel, pentacarbonyliron, and hexacarbonylchromium highly toxic volatile compounds, the most stable of which have filled valence shells of 18 electrons, (2) salts of anionic and cationic carbonyls, such as tetracarbonylcobaltate(−1) and hexacarbonylmanganese(+1), (3) dinuclear and polynuclear carbonyls, such as bis(tetracarbonylcobalt), the structural formula of which was shown earlier, and (4) mixed complexes containing other ligands in addition to CO: pentacarbonylchloromanganese, tetracarbonylhydridocobalt, and tricarbonylnitrosylcobalt.

Although molecular nitrogen, N_2, is isoelectronic with carbon monoxide (that is, it has the same number and arrangement of electrons), its tendency to form complexes with metals is much smaller. The first complex containing molecular nitrogen as a ligand—i.e., pentaamminenitrogenruthenium(2+), $[Ru(NH_3)_5(N_2)]^{2+}$ was prepared in 1965, and many others have been discovered subsequently. Such complexes have attracted considerable interest because of their possible roles in the chemical and biological fixation of nitrogen.

Nitrosyl Complexes

Nitrosyl complexes can be formed by the reaction of nitric oxide (NO) with many transition metal compounds or by reactions involving species containing nitrogen and oxygen. Some of these complexes have been known for many years—e.g., pentaaquanitrosyliron(2+) ion, $[Fe(H_2O)_5NO]^{2+}$, which formed in the classical brown-ring test for the qualitative detection of nitrate ion; Roussin's red ($K_2[Fe_2S_2(NO)_4]$) and black ($K[Fe_4S_3(NO)_7]$) salts; and sodium pentacyanonitrosylferrate(3−) dihydrate (sodium nitroprusside), $Na_2[Fe(CN)_5NO] \cdot 2H_2O$. Such complexes, which can be cationic, neutral, or anionic and which are usually deeply coloured (red, brown, purple, or black), have been extensively studied because they pose unique problems of structure and bonding and because they have potential uses as homogeneous catalysts for a variety of reactions. More recently, the research field has been expanded to include organometallic species.

Because the nitrosonium ion (NO^+) is isoelectronic with carbon monoxide and because its mode of coordination to transition metals is potentially similar to that of carbon monoxide, metal nitrosyls have been recognized as similar to carbonyls and are sometimes formulated as NO^+ complexes. Carbonyl ligands can be replaced by nitric oxide in substitution reactions. Such similarities may be deceptive, however, for the additional electron in neutral nitric oxide requires a more complicated treatment of M-NO bond formation. The NO ligand exhibits several geometries of coordination—linear (e.g., $[IrH(NO)\{P(C_6H_5)_3\}_3]^+$, $[Mn(CO)_2(NO)\{P(C_6H_5)_3\}_3]$, and $Na_2[Ru(OH)(NO_2)_4(NO)] 2H_2O$); bent (e.g., $[CoNO(NH_3)_5]^{2+}$ and $[IrCl_2(NO)\{P(C_6H_5)_3\}_2]$); or both (e.g., $[RuCl(NO)_2\{P(C_6H_5)_3\}_2]^+$). Like CO, NO also can act as a bridging ligand between two (e.g., $[\{Cr(\eta^5-C_5H_5)(NO)\}_2(\mu_2-NH_2)(\mu_2-NO)]$) or three (e.g., $[Mn_3(h^5-C_5H_5)_3(m_2-NO)_3(m_3-NO)]$) metal atoms. (The η^5 indicates that five carbon atoms of the $C_5H_5^-$ group are bonded to the chromium atom.)

Cyano and Isocyano Complexes

Cyano complexes, such as Prussian blue, mentioned above, are among the oldest coordination compounds. In addition to being a pseudohalide, the CN^- ion is isoelectronic with CO, RCN, RNC, N_2, and NO^+ (R is an alkyl group), and metal carbonyls and cyanide complexes are structurally similar. Also, like CO, CN^- enters into π as well as σ bonding with transition metal atoms or ions. Cyano complexes are among the most stable transition metal complexes; the extreme toxicity of CN^- (like that of CO) is due to its irreversible formation of a strong complex with hemoglobin, which prevents oxygen from binding reversibly to hemoglobin, thereby prohibiting the transport and release of oxygen in the body. Similarly, the ability of CN^- to form very stable complexes with silver $(Au(CN)_2^-)$ and gold $Au(CN)_2$ is the basis for its use in the extraction and purification of these metals. As a monodentate ligand, CN^- coordinates (bonds) through carbon as the donor atom, but, as a didentate ligand, it usually coordinates at both ends (C and N) and acts as a bridging ligand $(-CN-)$ to form infinite linear (chain) polymers as in Prussian blue, $AgCN, AuCN, Zn(CN)_2$, and $Cd(CN)_2$.

The cyanide ion forms complexes with transition metals and with zinc, cadmium, and mercury, usually by substitution in aqueous solution with no change in oxidation state. The most important complexes are anionic with the formula $[M^{n+}(CN)_x]^{(x-n)-}$, where Mn^+ represents a transition metal ion. Examples are $[Ni(CN)_4]^{2-}$, $[Pt(CN)_4]^{2-}, [Fe(CN)_6]^{4- \text{ or } 3-}, [Co(CN)_6]^{3-}, [Pt(CN)_6]^{2-}$, and $[Mo(CN)_8]^{5-, 4-, \text{ or } 3-}$. The free anhydrous parent acids of many of these anions for example, $H_4[Fe(CN)_6]$ and $H_3[Rh(CN)_6]$ have been isolated.

Cyanide complexes exhibit a variety of coordination numbers and configurations. Metal ions with a d^{10} structure form linear complexes of coordination number 2 as, for example, $[M(CN)_2]^-$ (where M = Hg, Ag, or Au)—while the isoelectronic complexes $[Cu(CN)_4]^{3-}, [Ag(CN)_4]^{3-}, [Zn(CN)_4]^{2-}, [Cd(CN)_4]^{2-}$, and $[Hg(CN)_4]^{2-}$ are tetrahedral. All the hexacoordinate complexes are octahedral, while octacoordinate complexes are cubic, dodecahedral, or square antiprismatic. (The dodecahedron and square antiprism are two structures that can be obtained by distorting the simple cube.) For d^2, d^4, d^6, d^8, and d^{10} transition metal ions, the octa-, hepta-, hexa-, penta-, and tetracoordinate complexes, respectively, are species with maximum coordination number.

Mixed complexes of type $[M(CN)_5 X]n^-$ (where $X = H_2O, NH_3, CO, NO, H$, or a halogen) also exist. The cyanide ion has the ability to stabilize metal ions in low oxidation states (probably by accepting electron density into its π^* orbitals)—e.g., $[Ni(CN)_4]^{4-}$, which contains nickel in the formal oxidation state of zero. Cyanide complexes have figured prominently in numerous kinetic studies. For example, fast electron-transfer reactions between $[Fe(CN)_6]^{3-}$ and $[Fe(CN)_6]^{4-}$ and between $[Mo(CN)_8]^{3-}$ and $[Mo(CN)_8]^{4-}$ established the outer-sphere mechanism for redox reactions ; replacement of water in $[Co(CN)_5 H_2O]^{2-}$ established the dissociative mechanism for substitution at a Co(3+) ion.

Transition metals also form complexes with organic cyanides (RCN or ArCN, called nitriles) and organic isocyanides (RNC or ArNC, called isonitriles)—where R and Ar are alkyl and aryl groups, respectively—by reaction of a metal halide, carbonyl, or other complex with the nitrile or isonitrile, respectively. Nitriles and isonitriles appear to be stronger donors of σ electrons than carbon monoxide, but they are capable of extensive back acceptance of π electrons from metals in lower oxidation states—as in $Cr(CNR)_6$

or $Cr(CNAr)_6$ and $Ni(CNR)_4$ or $Ni(CNAr)_4$, which are analogous to the corresponding carbonyls $Cr(CO)_6$ and $Ni(CO)_4$, respectively. Although a bridging isonitrile group has been reported in $(p\text{-}C_5H_5)_2 Fe_2 (CO)_3 (CNC_6H_5)$, this type of bonding is unusual.

Organometallic Complexes

Organometallic complexes are complexes formed between organic groups and metal atoms. They can be divided into two general classes: (1) complexes containing metal-carbon σ bonds and (2) π-bonded metal complexes of unsaturated hydrocarbons—that is, compounds with multiple bonds between carbon atoms.

Isopoly and Heteropoly Anions

The amphoteric metals of groups VB (vanadium, niobium, and tantalum) and VIB (chromium, molybdenum, and tungsten) in the +5 and +6 oxidation states, respectively, form weak acids that readily condense (polymerize) to form anions containing several molecules of the acid anhydride. If these condensed acids contain only one type of acid anhydride, they are called isopoly acids, and their salts are called isopoly salts. The acid anhydrides also can condense with other acids (e.g., phosphoric or silicic acids) to form heteropoly acids, which can form heteropoly salts. The condensation reactions, which occur reversibly in dilute aqueous solution, involve formation of oxo bridges by elimination of water from two molecules of the weak acid. The best-known and simplest example is the condensation of yellow chromate ion (CrO_4^{2-}) to form the orange isopoly dichromate ion ($Cr_2O_7^{2-}$), an equilibrium reaction the extent of which depends on the pH. In acidic solution the isopoly anion $Cr_2O_7^{2-}$, predominates while in basic solution the simple ion CrO_4^{2-} predominates.

$$2\,CrO_4^{2-} + H^+ \rightleftharpoons [O_3Cr\text{-}O\text{-}OH]^- \rightleftharpoons [O_3Cr\text{-}O\text{-}CrO_3]^{2-}\ H_2O$$

basic solution   acidic solution

Heteropoly acids and their salts may be formed by coordination of the central atom with four to six oxo anions, which may be mononuclear (containing one metal ion each), as in $H_7[P(MoO_4)_6]$, or trinuclear (containing three metal ions each), as in $H_3[P(W_3O_{10})_4]$. Incomplete replacement of oxygen atoms in PO_4^{3-} ions by MoO_3 groups can result in dimers (two-molecule polymers), as, for example, $\{OP[O(MoO_3)_3]_3\}_2^{6-}$. About 70 elements can act as central (hetero) atoms in heteropoly anions. Because each element may form more than one heteropoly anion and some of these anions can contain several different heteroatoms, thousands of heteropoly acids exist. Heteroatoms may be primary (these atoms are essential to the polyanion structure and thus not susceptible to chemical exchange) or secondary (these atoms can be removed by chemical reaction from the polyanion structure without destroying it). Heteropoly anions can be regarded as coordination compounds with polyanion ligands; e.g., $[(H_3N)_5 Cr(OH_2)]^{3+}$ can be considered the parent of $[(SiW_{11}O_{39})Cr(OH_2)]^{5-}$.

A variety of synthetic procedures are available for the preparation of isopoly acids and salts, which are usually less stable than heteropoly compounds. Heteropolymolybdates and heteropolytungstates are always prepared in solution, usually after acidifying and heating the theoretical amounts of reactants. In general, free heteropoly acids and salts, of which the heteropolymolybdates and heteropolytungstates are the best known, have very high molecular weights (some above 4,000) as compared with other inorganic electrolytes, are very soluble in water and organic solvents, are almost always highly hydrated with several hydrates existing, and are highly coloured. Some are strong oxidizing agents that can be reduced to stable, intensely deep blue species (heteropoly blues), which in turn can act as reducing agents, restoring the original colour on oxidation. The stoichiometry, oxidation-reduction potentials, and other characteristics of these reactions have been investigated by various methods. The free acids, which are polyprotic (contain several replaceable hydrogen ions), are fairly strong and nearly always stronger than the corresponding acids from which they are derived.

All heteropolymolybdate and heteropolytungstate anions are decomposed in strongly basic solution to form simple molybdate or tungstate ions and either an oxy anion or a hydrous metal oxide of the central metal atom, e.g.:

$$[P_2Mo_{18}O_{62}]^{6-} + 34\,OH^- \rightarrow 18\,MoO_4^{2-} + 2\,HPO_4^{2-} + 16\,H_2O \qquad [NiW_6O_{24}H_6]^{4-} + 8\,OH^- \rightarrow 6\,WO_4^{2-} + Ni(OH)_2 + 6\,H_2O$$

Throughout specific ranges of pH and other conditions, most solutions of heteropolymolybdates and heteropolytungstates appear to contain predominantly one distinct species of anion, many of which are remarkably stable and nonlabile.

The first heteropoly compound, $(NH_4)_3[PMo_{12}O_{40}]$, was obtained by the Swedish chemist Jöns Jacob Berzelius in 1826 as a yellow, crystalline precipitate, the formation of which is still used for the classical qualitative detection and quantitative estimation of phosphorus(after conversion to phosphate). By the beginning of the 20th century, hundreds of isopoly and heteropoly compounds were reported, many of which were based on incorrect analyses or failure to detect mixed crystals. Formulas were reported in terms of the old Berzelius dualistic theory as a combination of oxides, such as $3Na_2O \cdot Cr_2O_3 \cdot 12MoO_3 \cdot 20H_2O$ for $Na_3CrMo_6O_{24}H_6 \cdot 7H_2O$, and often merely expressed analytical results rather than structure. In addition to their use in analytical chemistry, heteropoly compounds have found use as catalysts, molecular sieves, corrosion inhibitors, photographic fixing agents, and precipitants for basic dyes.

Few structural studies of such compounds were carried out, but this lack did not prevent the elaboration of various unsuccessful theories to account for their structures. In 1907 Werner applied his coordination theory to the structure of 12-tungstosilicic acid, $H_4[SiW_{12}O_{40}]$, and its salts by assuming that the central group is an SiO_4^{4-} ion surrounded octahedrally by six $RW_2O_6^+$ groups (R = a unipositive ion), four linked by primary (ionic) and two linked by secondary (coordinate covalent) valences. Difficulties were encountered by this system as well as by the later (1910–21), more elaborate Miolati-Rosenheim theory. Modern conclusive knowledge of the structures of heteropoly compounds did not begin until 1934, with J.F. Keggin's determination of the structure of $PO_4W_{12}O_{36}]\cdot5H_2O$ by the most direct means, X-ray diffraction.

The structures of isopoly and heteropoly compounds consist of polyhedrons sharing corners and edges with one another. In heteropolymolybdates or heteropolytungstates, each molybdenum or tungsten atom is located at the centre of an octahedron, each vertex of which is occupied by an oxygen atom. These octahedrons can share corners or edges or both with other MoO_6 or WO_6 octahedrons. In $[Mo_8O_{26}]^{4-}$ eight MoO_6 octahedrons share edges. In $[PMo_{12}O_{40}]^{3-}$ the central phosphorus atom is located at the centre of a PO_4 tetrahedron, which is surrounded by 12 MoO_6 octahedrons, which share corners so that the correct number of oxygen atoms is utilized.

In coordination chemistry, the coordination number is the number of ligands attached to the central ion (more specifically, the number of donor atoms). Coordination num-bers are normally between two and nine. The number of bonds depends on the size, charge, and electron configuration of the metal ion and the ligands.

Typically the chemistry of complexes is dominated by interactions between s and p molecular orbitals of the ligands and the d orbitals of the metal ions. The s, p, and d orbitals of the metal can accommodate 18 electrons. The maximum coordination number for a certain metal is thus related to the electronic configuration of the metal ion (specifically, the number of empty orbitals) and to the ratio of the size of the ligands and the metal ion. Large metals and small ligands lead to high coordination numbers (e.g., $[Mo(CN)_8]^{4-}$). Small metals with large ligands lead to low coordination numbers (e.g., $Pt[P(CMe_3)]_2$). Due to their large size, lanthanides, actinides, and early transition metals tend to have high coordination numbers.

Coordination Number Examples

- Carbon has a coordination number of 4 in methane (CH_4) molecule since it has four hydrogen atoms bonded to it.

- In ethylene ($H_2C=CH_2$), the coordination number of each carbon is 3, where each C is bonded to 2H + 1C for a total of 3 atoms.

- The coordination number of diamond is 4, as each carbon atom rests at the center of a regular tetrahedron formed by four carbon atoms.

Calculating the Coordination Number

Here are the steps for identifying the coordination number of a coordination compound.

1. Identify the central atom in the chemical formula. Usually, this is a transition metal.

2. Locate the atom, molecule, or ion nearest the central metal atom. To do this, find the molecule or ion directly beside the metal symbol in the chemical formula of the coordination compound. If the central atom is in the middle of the formula, there will be neighboring atoms/molecules/ions on both sides.

3. Add the number of atoms of the nearest atom/molecule/ions. The central atom may only be bonded to one other element, but you still need to note the number of atoms of that element in the formula. If the central atom is in the middle of the formula, you'll need to add up the atoms in the entire molecule.

4. Find the total number of nearest atoms. If the metal has two bonded atoms, add together both numbers.

Ligands

In coordination chemistry, a ligand is an ion or molecule (functional group) that binds to a central metal atom to form a coordination complex. Virtually every molecule and every ion can serve as a ligand for (or coordinate to) metals. Denticity refers to the number of times a ligand bonds to a metal through donor atoms. Many ligands are capable of binding metal ions through multiple sites, usually because the ligands have lone pairs on more than one atom.

Monodentate ligands include virtually all anions and all simple Lewis bases. Thus, the halides and pseudohalides are important anionic ligands. Ammonia, carbon monoxide, and water are particularly common charge-neutral ligands. Simple organic species are also very common. All unsaturated molecules are also ligands, utilizing their π-electrons in forming the coordinate bond. Also, metals can bind to the σ bonds in, for example, silanes, hydrocarbons, and dihydrogen.

Ligands that bind via more than one atom are often termed polydentate or chelating. A ligand that binds through two sites is classified as bidentate, and three sites as tridentate. Chelating ligands are commonly formed by linking donor groups via organic linkers. A classic bidentate ligand is ethylenediamine, which is derived by the linking of two ammonia groups with an ethylene ($-CH_2CH_2-$) linker. A classic example of a polydentate ligand is the hexadentate chelating agent EDTA, which is able to bond through six sites, completely surrounding some metals.

There are several types of polydentate ligands which can be characterized based on how they interact with the central ion. For example, trans-spanning ligands are bidentate ligands that can span coordination positions on opposite sides of a coordination complex. Ambidentate ligands can attach to the central atom in two places but not both. A bridging ligand links two or more metal centers. Changing the size and electronic properties of ligands can be used to control catalysis of the central ion and stabilize unusual coordination sites.

Geometries

Different ligand structural arrangements result from the coordination number. Most structures follow the pattern as if the central atom were in the middle and the corners of that shape are the locations of the ligands. These shapes are defined by orbital overlap between ligand and metal orbitals and ligand-ligand repulsions, which tend to lead to certain regular geometries. However, there are many cases that deviate from regular geometry. For example, ligands of different sizes and with different electronic effects often result in irregular bond lengths.

Coordination Number 2

This arrangement is not very common for first row transition metal ion complexes and some of the best known examples are for Silver(I). In this case we have a low charge and an ion at the right hand side of the d-block indicating smaller size.

A method that was often employed for the detection of chloride ions involved the formation of the linear diamminesilver(I) complex.

The first step is:

$$Ag^+ + Cl^- \rightarrow AgCl\,(\text{white ppt})$$

and to ensure that the precipitate is really the chloride salt, two further tests must be done:

$$AgCl + 2\,NH_3 \rightarrow \left[Ag(NH_3)_2\right]+$$

and

$$\left[Ag(NH_3)_2\right]^+ + HNO_3 \rightarrow AgCl\,(\text{re-ppts})$$

The reaction of a bidentate ligand such as 1,2-diaminoethane with Ag(I) does not lead to chelated ring systems, but instead to linear two coordinate complexes. One reason for this is that bidentate ligands can NOT exist in *trans* arrangements, that is they are UNABLE to span 180 degrees.

Coordination Number 3

Once again, this is not very common for first row transition metal ions. Examples with three different geometries have been identified:

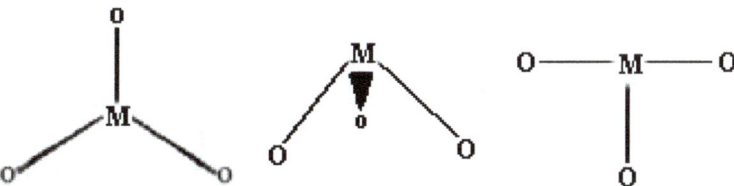

Trigonal Planar

Well known for main group species like CO_3^{2-} etc., this geometry has the four atoms in a plane with the bond angles between the ligands at 120 degrees.

Trigonal Pyramid

More common with main group ions.

T-shaped

The first example of a rare T-shaped molecule was found in 1977 however since then several further examples have been reported.

Coordination Number 4

Two different geometries are possible. The tetrahedron is the more common while the square planar is found in particular with metal ions having a d^8 electronic configuration.

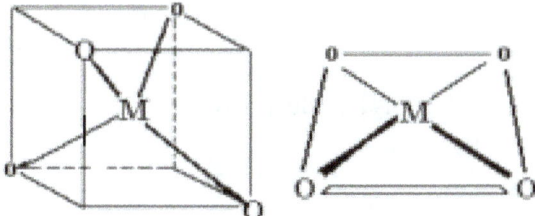

Tetrahedral, (T_d)

The chemistry of molecules centred around a tetrahedral C atom is covered in organic courses. To be politically correct, please change all occurrences of *C* to *Co*. There is a large number of tetrahedral cobalt(II) complexes known.

Square Planar, (D_4h)

These are much less common than tetrahedral and are included here since there are some extremely important examples with this shape.

Coordination Number 5

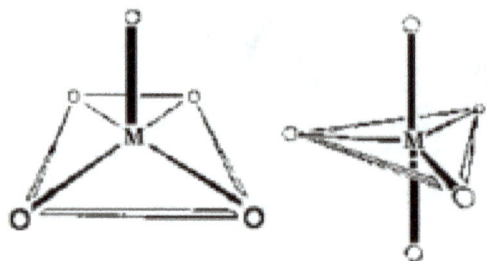

Square Pyramid, (C_{4v})

Oxovanadium salts (Vanadyl, VO^{2+}) often show square pyramidal geometry, for example, $VO(acac)_2$. Note that the Vanadium(IV) can be considered coordinatively unsaturated and addition of pyridine leads to the formation of an octahedral complex.

Trigonal Bipyramid, (D_{3h})

The structure of $[Cr(en)_3][Ni(CN)_5]_{1.5}H_2O$ was reported in 1968 to be a remarkable example of a complex exhibiting both types of geometry in the same crystal. The reaction of cyanide ion with Ni^{2+} proceeds via several steps:

$$Ni^{2+} + 2\,CN^- \;\circledR\; Ni(CN)_2 \;\; yellow$$

$$Ni(CN)_2 + 2\,CN^- \rightarrow \left[Ni(CN)_4\right]^{2-} \;\; orange\text{-}red$$

$$\log(\beta 4) = 30.1$$

$$\left[Ni(CN)_4\right]^{2-} + CN^- \rightarrow \left[Ni(CN)_5\right]^{3-} \;\; deep\ red$$

Coordination Number 6

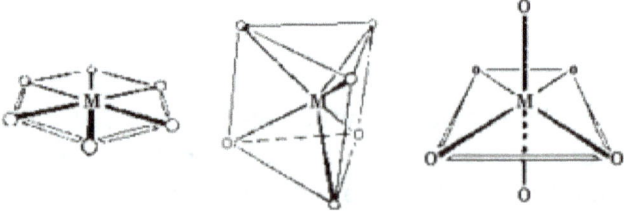

Hexagonal Planar

Unknown for first row transition metal ions, although the arrangement of six groups in a plane is found in some higher coordination number geometries.

Trigonal Prism

Most trigonal prismatic compounds have three bidentate ligands such as dithiolates or oxalates and few are known for first row transition metal ions.

Octahedral, (O_h)

The most common geometry found for first row transition metal ions, including all aqua ions. In some cases distortions are observed and these can sometimes be explained in terms of the Jahn-Teller Theorem.

Coordination Number 7

Three geometries are possible:

Not very common for 1st row complexes and the energy difference between the structures seems small and distortions occur so that prediction of the closest "idealised" shape is generally difficult.

 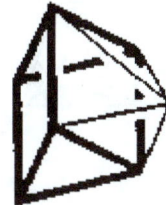 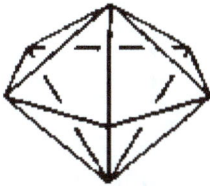

Capped octahedron, (C_{3v})

Capped trigonal prism, (C_{2v})

Pentagonal Bipyramid, (D_{5h})

Coordination Number 8

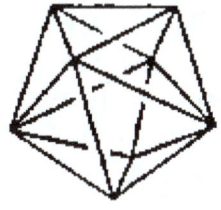 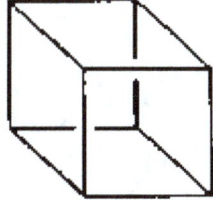 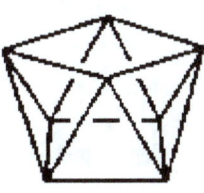

Dodecahedron, (D_{2d})

Cube, (O_h)

Square antiprism, (D_{4d})

Hexagonal bipyramid, (D_{6h})

Coordination Number 9

Three-face centred trigonal prism, (D_{3h})

Coordination Number 10

Bicapped square antiprism, (D_{4d})

Coordination Number 11

All-faced capped trigonal prism, (D_{3h})

Coordination Number 12

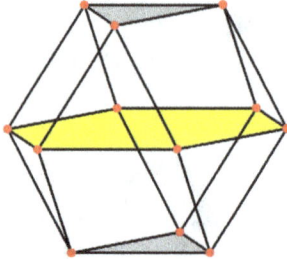

cuboctahedron, (O_h)

Coordination Sphere

In a complex compound, it usually, central metal ion and the ligands are enclosed within square bracket is called a coordination sphere. This represents a single constituent unit. The ionizable species are placed outside the square bracket.

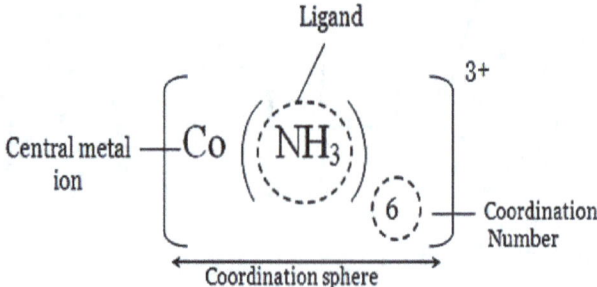

First Coordination Sphere

The first coordination sphere refers to the molecules that are attached directly to the metal. These molecules are typically solvent. The interactions between the first and second coordination spheres usually involve hydrogen-bonding. For charged complexes, ion pairing is important.

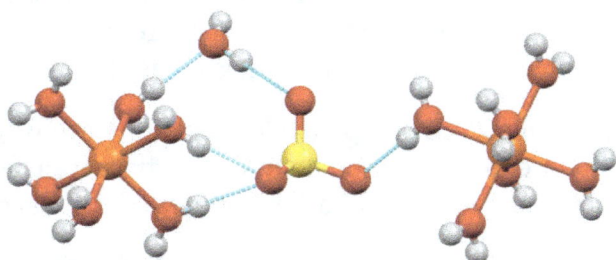

Hexamminecobalt(III) chloride is a salt of a coordination complex wherein six ammonia ("ammine") ligands occupy the first coordination sphere of the ion Co^{3+}.

In hexamminecobalt(III) chloride ($[Co(NH_3)_6]Cl_3$), the cobalt cation plus the 6 ammonia ligands comprise the first coordination sphere. The coordination sphere of this ion thus consists of a central MN_6 core "decorated" by 18 N−H bonds that radiate outwards.

Second Coordination Sphere

In crystalline $FeSO_4 \cdot 7H_2O$, the *first* coordination sphere of Fe^{2+} consists of six water ligands. The *second* coordination sphere consists of a water of crystallization and sulfate, which interact with the $[Fe(H_2O)_6]^{2+}$ centers.

Metal ions can be described as consisting of series of two concentric coordination spheres, the first and second. More distant from the second coordination sphere, the solvent molecules behave more like "bulk solvent." Simulation of the second coordination sphere is of interest in computational chemistry. The second coordination sphere can consist of ions (especially in charged complexes), molecules (especially those that hydrogen bond to ligands in the first coordination sphere) and portions of a ligand backbone. Compared to the first coordination sphere, the second coordination sphere has a less direct influence on the reactivity and chemical properties of the metal complex. Nonetheless the second coordination sphere is relevant to understanding reactions of the metal complex, including the mechanisms of ligand exchange and catalysis.

Role in Catalysis

Mechanisms of metalloproteins often invoke modulation of the second coordination sphere by the protein. For example, an amine cofactor in the second coordination sphere of some hydrogenase enzymes assists in the activation of dihydrogen substrate.

In metal complexes of 1,5-diaza-3,7-diphosphacyclooctanes and related ligands, amine groups occupy the second coordination sphere.

Role in Mechanistic Inorganic Chemistry

The rates at which ligands exchange between the first and the second coordination sphere is the first step in ligand substitution reactions. In associative ligand substitution, the entering nucleophile resides in the second coordination sphere. These effects are relevant to practical applications such as contrast agents used in MRI.

The energetics of inner sphere electron transfer reactions are discussed in terms of second coordination sphere. Some proton coupled electron transfer reactions involve atom transfer between the second coordination spheres of the reactants:

$$[Fe*(H_2O)_6]^{2+} + [Fe(H_2O)_5(OH)]^{2+} \quad [Fe(H_2O)_6]^{3+} + [Fe*(H_2O)_5(OH)]^{2+}$$

Role in Spectroscopy

Solvent effects on colors and stability are often attributable to changes in the second coordination sphere. Such effects can be pronounced in complexes where the ligands in the first coordination sphere are strong hydrogen-bond donors and acceptors, e.g. respectively $[Co(NH_3)_6]^{3+}$ and $[Fe(CN)_6]^{3-}$. Crown-ethers bind to polyamine complexes through their second coordination sphere. Polyammonium cations bind to the nitrogen centres of cyanometallates.

Role in Supramolecular Chemistry

Macrocyclic molecules such as cyclodextrins act often as the second coordination sphere for metal complexes.

Trans and Cis Effect

The *trans* effect proper, which is often called the kinetic *trans* effect, refers to the observation that certain ligands increase the rate of ligand substitution when positioned *trans* to the departing ligand. The key word in that last sentence is "rate"—the *trans* effect proper is a kinetic effect. The *trans* influence refers to the impact of a ligand on the length of the bond *trans* to it in the ground state of a complex. The key phrase there is "ground state"—this is a thermodynamic effect, so it's sometimes called the thermodynamic *trans* effect. Adding to the insanity, *cis* effects and *cis* influences have also been observed. Evidently, ligands may also influence the kinetics or thermodynamics of their *cis* neighbors. All of these phenomena are independent of the metal center, but *do* depend profoundly on the geometry of the metal (more on that shortly).

Kinetic *trans* and *cis* effects are shown in the figure below. In both cases, we see that X^1 exhibits a stronger effect than X^2. The geometries shown are those for which each effect is most commonly observed. The metals and oxidation states shown are prototypical.

The kinetic *trans* and *cis* effects in action. X^1 is the stronger (*trans/cis*)-effect ligand in these examples.

On to the influences, which are simpler to illustrate since they're concerned with ground states, not reactions. The lengthened bonds below are exaggerated.

The *trans* and *cis* influences in action. Note the elongated bond lengths.

And there we have it folks, the thermodynamic and kinetic *cis/trans* effects. It's worth keeping in mind that the kinetic *trans* effect is most common for d^8 square planar complexes, and the kinetic *cis* effect is most common for d^6 octahedral complexes (particularly when the departing L is CO). But a lingering question remains: what makes for a strong *trans* effect ligand?

Origins of Effects & Influences

The *trans* effect and its cousins are all electronic, not steric effects. So, the electronic properties of the ligand dictate the strength of its *trans* effect. Let's finally dig into the *trans* effect series:

$$(weak)\ F^-,\ HO^-,\ H_2O < NH_3 < py\ <\ Cl^- < Br^- < I^-,\ SCN^-,\ NO_2^-,\ SC(NH_2)_2,$$
$$Ph^- <\ SO_3^{2-} < PR_3 < AsR_3,\ SR_2,\ H_3C^- < H^-,\ NO,\ CO,\ NC^-,\ C_2H_4\ (strong)$$

What's the electronic progression here? It's clear that electronegativity decreases across the series: $F^- < Cl^- < Br^- < I^- < H_3C^-$. From a bonding perspective, we can say that ligands with strong *trans* effects are strong σ-donors (or σ-bases). Yet σ-donation doesn't tell the whole story. What about ethylene and carbon monoxide, which both appear at the top of the heap? Neither of these ligands are strong σ-donors, but their π systems do interact with the metal center through backbonding. Consider the following sub-series: $S = C = N^- < PR_3 < CO$. Backbonding increases across this series, along with the strength of the *trans* effect. Strong backbonders—better known as π-acceptors or π-acids—exhibit strong *trans* effects.

Strong trans effect = strong s- donor + strong p- acceptor

Wonderful! Using these ideas we can identify ligands with strong *trans* effects. But we can dive deeper down the rabbit hole: *why* does this particular combination of electronic factors lead to a strong *trans* effect? To understand this, we need to know the mechanism of the ligand substitution reaction that's sped up by strong *trans* effect ligands. For 16-electron Pt(II) complexes, associative substitution is par for the course. The incoming ligand binds to the metal first, forming an 18-electron complex (yay!), which jettisons a ligand to yield a new 16-electron product. The mechanism in all its glory is shown in the figure below.

The mechanism of associative ligand substitution of Pt(II) complexes.

Some very important points about this mechanism:

- The incoming ligand always sits at an equatorial site in the trigonal bipyramidal intermediate. More on this another day, but I think of this result as governed by the principle of least motion. Consider the molecular gymnastics that would have to happen to place the incoming ligand in an axial position.

- Two ligands in the square plane are "pushed down" and become the other two equatorial ligands.

- Owing to microscopic reversibility, the leaving group must be one of the equatorial ligands.

The third point reveals that once L' has "pushed down" X^{TE} and L^{trans}, L^{trans} has no choice but to leave (assuming X^{TE} stays put). Thus, the *trans* effect has nothing to do with the second step of the mechanism, which is not rate determining anyway. The key is the first step—in particular, the "pushing down" event. Apparently, ligands with strong *trans* effects like to be pushed down. They like to occupy the equatorial plane of the TBP intermediate. Now here's the kicker: the equatorial *sites* of the TBP geometry are more π *basic* than the axial sites. The equatorial plane is just the xy-plane of the metal center, and the d orbitals in that plane (when occupied) are great electron sources for π-acidic ligands. Thus, π-acidic ligands want to occupy those equatorial sites, to receive the benefits of strong backbonding! Boom; strong π-acids encourage loss of the ligand *trans* to themselves.

The equatorial sites of TBP metals are rich in electrons that can π bond.

What about those pesky σ donors? Well, we can imagine that in a square planar complex, a ligand and its *trans* partner are competing for donation into the same d orbital. Strong σ donation from a ligand should thus weaken the bond *trans* to it. Although this is the thermodynamic *trans* effect (*trans* influence) in action, the resulting destabilization of the ground state relative to the transition state is a kinetic effect. On the whole, the barrier to substitution of the *trans* ligand goes down as σ-donating strength goes up.

This idea of "competition for the metal center" is a nice heuristic to use when thinking about the *trans* and *cis*influences. The type of metallic orbital involved in M−L bonding determines the strength of L's *trans* and *cis*influences on neighboring ligands that also need that metallic orbital for bonding. For example: both influences are large if the metal's s orbital is a significant contributor to M−L bonding, since it's non-directional; the *trans* influence is much greater than the *cis* influence when metallic p orbitals are primarily involved in M−L.

Jahn–Teller Effect

The Jahn-Teller effect is a geometric distortion of a non-linear molecular system that reduces its symmetry and energy. This distortion is typically observed among octahedral complexes where the two axial bonds can be shorter or longer than those of the equatorial bonds. This effect can also be observed in tetrahedral compounds. This effect is dependent on the electronic state of the system.

In 1937, Hermann Jahn and Edward Teller postulated a theorem stating that "stability and degeneracy are not possible simultaneously unless the molecule is a linear one," in regards to its electronic state. This leads to a break in degeneracy which stabilizes the molecule and by consequence, reduces its symmetry. Since 1937, the theorem has been revised which Housecroft and Sharpe have eloquently phrased as "any non-linear molecular system in a degenerate electronic state will be unstable and will undergo distortion to form a system of lower symmetry and lower energy, thereby removing the degeneracy." This is most commonly observed with transition metal octahedral complexes, however, it can be observed in tetrahedral compounds as well.

For a given octahedral complex, the five d atomic orbitals are split into two degenerate sets when constructing a molecular orbital diagram. These are represented by the sets' symmetry labels: t_{2g} (d_{xz}, d_{yz}, d_{xy}) and e_g $(d_{z^2}$ and $d_{x^2-y^2})$. When a molecule possesses a degenerate electronic ground state, it will distort to remove the degeneracy and form a lower energy (and by consequence, lower symmetry) system. The octahedral complex will either elongate or compress the z ligand bonds as shown in Figure below:

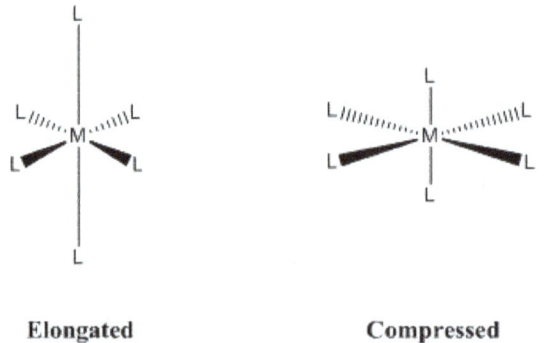

      Elongated   **Compressed**

Figure: Jahn-Teller distortions for an octahedral complex.

When an octahedral complex exhibits elongation, the axial bonds are longer than the equatorial bonds. For a compression, it is the reverse; the equatorial bonds are longer than the axial bonds. Elongation and compression effects are dictated by the amount of overlap between the metal and ligand orbitals. Thus, this distortion varies greatly depending on the type of metal and ligands. In general, the stronger the metal-ligand orbital interactions are, the greater the chance for a Jahn-Teller effect to be observed.

Elongation

Elongation Jahn-Teller distortions occur when the degeneracy is broken by the stabilization (lowering in energy) of the d orbitals with a z component, while the orbitals without a z component are destabilized (higher in energy) as shown in figure below:

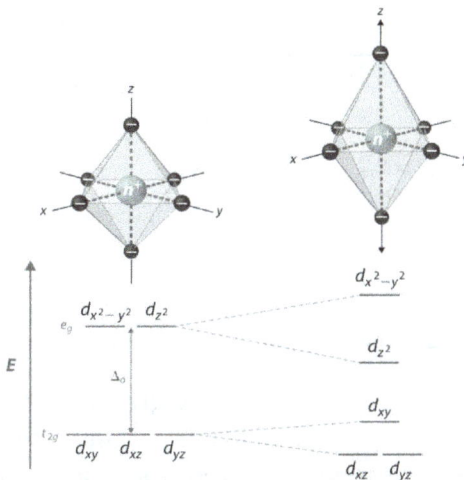

Figure: Illustration of tetragonal distortion (elongation) for an octahedral complex.

This is due to the d_{xy} and $d_{x^2-y^2}$ orbitals having greater overlap with the ligand orbitals, resulting in the orbitals being higher in energy. Since the $d_{x^2-y^2}$ orbital is antibonding, it is expected to increase in energy due to elongation. The d_{xy} orbital is still nonbonding, but is destabilized due to the interactions. Jahn-Teller elongations are well-documented for copper(II) octahedral compounds. A classic example is that of copper(II) fluoride as shown in figure below.

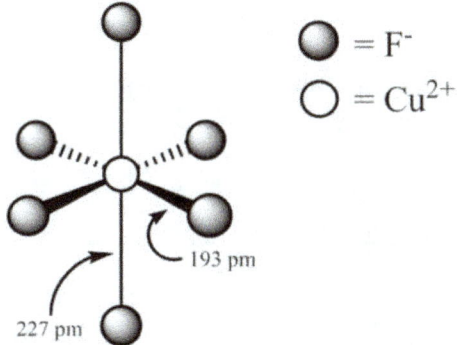

$\bullet$ = F⁻
$\bigcirc$ = Cu²⁺

Figure: Structure of octahedral copper(II) fluoride.

Notice that the two axial bonds are both elongated and the four shorter equatorial bonds are the same length as each other. According the theorem, the orbital degeneracy is eliminated by distortion, making the molecule more stable based on the model presented in Figure.

Compression

Compression Jahn-Teller distortions occur when the degeneracy is broken by the stabilization (lowering in energy) of the d orbitals *without* a z component, while the orbitals with a z component are destabilized (higher in energy) as shown in figure below:

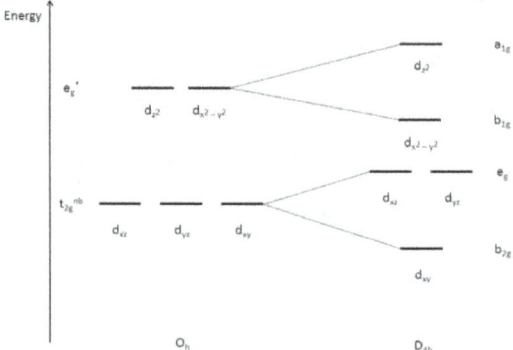

Figure: Illustration of tetragonal distortion (compression) for an octahedral complex.

This is due to the z-component d orbitals having greater overlap with the ligand orbitals, resulting in the orbitals being higher in energy. Since the d_{z^2} orbital is antibonding, it is expected to increase in energy due to compression. The d_{xz} and d_{yz} orbitals are still nonbonding, but are destabilized due to the interactions.

Electronic Configurations

For Jahn-Teller effects to occur in transition metals there must be degeneracy in either the t_{2g} or e_g orbitals. The electronic states of octahedral complexes are classified as either low spin or high spin. The spin of the system is dictated by the chemical environment. This includes the characteristics of the metal center and the types of ligands.

Low Spin

Figure (below) shows the various electronic configurations for low spin octahedral complexes:

Figure: Low spin octahedral coordination diagram (red indicates no degeneracies possible, thus no Jahn-Teller effects).

The figure illustrates that low spin complexes with d^3, d^5, d^8, and d^{10} electrons that do no exhibit Jahn-Teller distortions. These electronic configurations correspond to a variety of transition metals. Many common examples include Cr^{3+}, Co^{3+}, and Ni^{2+}.

High Spin

Figure shows the various electronic configurations for high spin octahedral complexes:

Figure: High spin octahedral coordination diagram (red indicates no degeneracies possible, thus no Jahn-Teller effects).

The figure illustrates that low spin complexes with d^3, d^5, d^8, and d^{10} electrons cannot have Jahn-Teller distortions. In general, degenerate electronic states occupying the egeg orbital set tend to show stronger Jahn-Teller effects. This is primarily caused by the occupation of these high energy orbitals. Since the system is more stable with a lower energy configuration, the degeneracy of the eg set is broken, the symmetry is reduced, and occupations at lower energy orbitals occur.

Spectroscopic Observation

Jahn-Teller distortions can be observed using a variety of spectroscopic techniques. In UV-VIS absorption spectroscopy, distortion causes splitting of bands in the spectrum due to a reduction in symmetry (O_h to D_{4h}). Consider a hypothetical molecule with octahedral symmetry showing a single absorption band. If the molecule were to undergo Jahn-Teller distortion, the number of bands would increase as shown in figure below:

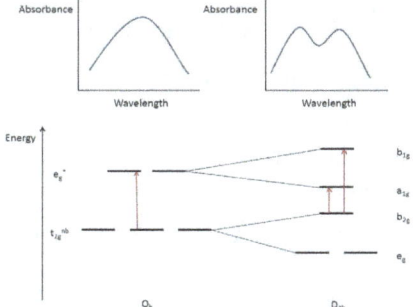

Figure: Hypothetical absorption spectra of an octahedral molecule (left) and the same molecule with Jahn-Teller elongation (right). The red arrows indicate electronic transitions.

A similar phenomenon can be seen with IR and Raman vibrational spectroscopy. The number of vibrational modes for a molecule can be calculated using the 3n - 6 rule (or 3n - 5 for linear geometry) rule. If a molecule exhibits an O_h symmetry point group, it will have fewer bands than that of a Jahn-Teller distorted molecule with D_{4h} symmetry. Thus, one could observe Jahn-Teller effects through either IR or Raman techniques. This effect can also be observed in EPR experiments as long as there is at least one unpaired electron.

Table: *Examples of Jahn-Teller distorted complexes*	
$CuBr_2$	4 Br at 240 pm 2 Br at 318 pm
$CuCl_2$	4 Cl at 230 pm 2 Cl at 295 pm
$CuCl_2.2H_2O$	2 O at 193 pm 2 Cl at 228 pm 2 Cl at 295 pm
$CsCuCl_3$	4 Cl at 230 pm 2 Cl at 265 pm
CuF_2	4 F at 193 pm 2 F at 227 pm
$CuSO_4.4NH_3.H_2O$	4 N at 205 pm 1 O at 259 pm 1 O at 337 pm
K_2CuF_4	4 F at 191 pm 2 F at 237 pm
$KCuAlF_6$	2 F at 188 pm 4 F at 220 pm
CrF_2	4 F at 200 pm 2 F at 243 pm
$KCrF_3$	4 F at 214 pm 2 F at 200 pm
MnF_3	2 F at 209 pm 2 F at 191 pm 2 F at 179 pm

The Jahn-Teller Theorem predicts that distortions should occur for *any degenerate state*, including degeneracy of the t_{2g} level, however distortions in bond lengths are much more distinctive when the degenerate electrons are in the e_g level.

References

- Coordination-compound, science: britannica.com: Retrieved 16 April 2018

- Ligand-field-and-molecular-orbital-theories-277802: britannica.com, Retrieved 19 June 2018

- J. C. Fontecilla-Camps, A. Volbeda, C. Cavazza, Y. Nicolet "Structure/Function Relationships of [NiFe]- and [FeFe]-Hydrogenases" Chem. Rev. 2007, 107, 4273-4303. doi:10.1021/cr050195z

- Coordination-number-ligands-and-geometries: courses.lumenlearning.com, Retrieved 24 July 2018

- Definition-of-coordination-number-604956: thoughtco.com, Retrieved 09 April 2018

- What-is-coordination-sphere, chemistry: qsstudy.com, Retrieved 16 April 2018

- R. M. Supkowski, W. DeW. Horrocks Jr. "On the determination of the number of water molecules, q, coordinated to europium(III) ions in solution from luminescence decay lifetimes" Inorganic Chimica Acta 2002, Volume 340, pp. 44–48. doi:10.1016/S0020-1693(02)01022-8

- The-transcis-effects-influences: organometallicchem.wordpress.com, Retrieved 31 March 2018

Permissions

All chapters in this book are published with permission under the Creative Commons Attribution Share Alike License or equivalent. Every chapter published in this book has been scrutinized by our experts. Their significance has been extensively debated. The topics covered herein carry significant information for a comprehensive understanding. They may even be implemented as practical applications or may be referred to as a beginning point for further studies.

We would like to thank the editorial team for lending their expertise to make the book truly unique. They have played a crucial role in the development of this book. Without their invaluable contributions this book wouldn't have been possible. They have made vital efforts to compile up to date information on the varied aspects of this subject to make this book a valuable addition to the collection of many professionals and students.

This book was conceptualized with the vision of imparting up-to-date and integrated information in this field. To ensure the same, a matchless editorial board was set up. Every individual on the board went through rigorous rounds of assessment to prove their worth. After which they invested a large part of their time researching and compiling the most relevant data for our readers.

The editorial board has been involved in producing this book since its inception. They have spent rigorous hours researching and exploring the diverse topics which have resulted in the successful publishing of this book. They have passed on their knowledge of decades through this book. To expedite this challenging task, the publisher supported the team at every step. A small team of assistant editors was also appointed to further simplify the editing procedure and attain best results for the readers.

Apart from the editorial board, the designing team has also invested a significant amount of their time in understanding the subject and creating the most relevant covers. They scrutinized every image to scout for the most suitable representation of the subject and create an appropriate cover for the book.

The publishing team has been an ardent support to the editorial, designing and production team. Their endless efforts to recruit the best for this project, has resulted in the accomplishment of this book. They are a veteran in the field of academics and their pool of knowledge is as vast as their experience in printing. Their expertise and guidance has proved useful at every step. Their uncompromising quality standards have made this book an exceptional effort. Their encouragement from time to time has been an inspiration for everyone.

The publisher and the editorial board hope that this book will prove to be a valuable piece of knowledge for students, practitioners and scholars across the globe.

Index

www.ingramcontent.com/pod-product-compliance
Lightning Source LLC
Chambersburg PA
CBHW080402190526
45161CB00003B/111